T. H. Aldrich, Otto Meyer

Preliminary Report on the Tertiary Fossils of Alabama and Mississippi...

T. H. Aldrich, Otto Meyer

Preliminary Report on the Tertiary Fossils of Alabama and Mississippi...

ISBN/EAN: 9783337139704

Printed in Europe, USA, Canada, Australia, Japan

Cover: Foto ©berggeist007 / pixelio.de

More available books at **www.hansebooks.com**

GEOLOGICAL SURVEY

OF

ALABAMA.

EUGENE A. SMITH, Ph.D., State Geologist.

BULLETIN No. 1.

I.

PRELIMINARY REPORT ON THE TERTIARY FOSSILS OF ALABAMA AND MISSISSIPPI.

By TRUMAN H. ALDRICH, M. E.

II.

CONTRIBUTIONS TO THE EOCENE PALEONTOLOGY OF ALABAMA AND MISSISSIPPI,

By OTTO MEYER, Ph. D.

PRINTED FOR THE GEOLOGICAL SURVEY.

1886.

To His Excellency, E. A. O'Neal, Governor of Alabama.

Sir:—I have the honor to transmit herewith Bulletin No. 1 of the Geological Survey, which is the first contribution toward a work undertaken by Mr. Truman H. Aldrich, illustrating the Paleontology of the Tertiary Formation in Alabama. This work, which is to be the gift of Mr. Aldrich to the State of Alabama, will embrace figures and descriptions of all the shells found in the Tertiary deposits of the State, including reproductions of the figures already published elsewhere, and when finished, will be one of the most complete works of the kind published by any State.

In the preparation of this Bulletin, Mr. Aldrich has personally gone over the greater part of the ground, and has collected a large part of the material himself; he is thus able to give to each species, not only its locality, but also its exact place in the stratigraphical scale.

The work is therefore, not a bare description of new species, of no interest or value except to the paleontologist, but it illustrates very fully the distribution of the species both in time and space; and in order that these relations may be more clearly set before the reader, I have added an extract from my forthcoming Geological Report, giving a summary of the lithological and stratigraphical features and subdivisions of the various deposits which make up the Tertiary Formation of Alabama.

The lower or Jackson division of the White Limestone in the State of Mississippi contains fossils in a much better state of preservation than any as yet found at corresponding horizons in Alabama. For this reason, some of the figures and descriptions in the following Bulletin refer to shells actually collected in Mississippi, though probably also occurring within the limits of Alabama.

That part of the Bulletin which bears the name of Dr. Otto Meyer, including the drawing of the three plates illustrating his article, has been furnished by the author free of all charge, and our acknowledgments are hereby made to him for his valuable contribution.

Mr. Aldrich not only furnishes gratuitously the text of his article, but also provides at his own expense the entire printed edition of the plates illustrating both his own and Dr. Meyer's articles. The only cost to the State is for the printing of a few pages of ordinary letter-press.

In answer to the possible question as to what practical or economical value such work may have, it may be stated that there are at intervals through the Tertiary deposits of Alabama, ten or fifteen horizons at which occur beds of marine shells. In many cases these are closely associated with green-sands and other materials of economic value, the recognition of which becomes thus often a matter of importance.

The association of the shells in these marine deposits is always such that the different beds may in general be very easily identified and distinguished from each other by any one even moderately well acquainted with their fossil contents.

These shells thus become most valuable guides by which the geologist is enabled to recognize the identity of strata widely separated geographically, and differing widely in the character of the materials of which they are composed.

With the small appropriation available for the purpose of the survey, I should have hesitated to expend any of it upon a subject like this, which, while it has, as above shown, very important practical and economical bearings, might yet be looked upon as of comparatively little utility to the majority of the people of the State; therefore, our thanks are due to Mr. Aldrich for coming to our aid, both as a writer and financially.

EUGENE A. SMITH.

University of Alabama, *July* 1, 1886.

TABLE OF CONTENTS.

INTRODUCTION.

Summary of the Lithological and Stratigraphical Features and Subdivisions of the Tertiary of Alabama.

In the Report of the Geological Survey of Alabama for the years 1883-'4, these relations of the Tertiary Formation are described in detail and fully illustrated by plates of sections and a map. From this report the following summary is taken.

THICKNESS.

The whole thickness of the strata of the Tertiary Formation of Alabama, occurring in the vicinity of the two rivers, is between 1620 and 1700 feet. This estimate is based upon actual measurements, except at one or two horizons, and even in these places we are able to give close estimates of the thickness of the strata not measured.

SUBDIVISIONS.

We have adopted a fourfold division of the Tertiary, which, in descending order, is the following:

1. The White Limestone,
2. The Claiborne,
3. The Buhrstone and
4. The Lignitic.

In all that follows the strata are described in descending order.

1. *The White Limestone.*—This subdivision is calcareous throughout, but the lowermost 60 feet are more argillaceous than the rest. The

minimum thickness is 350 feet, of which the uppermost 150 feet consist of a tolerably pure but somewhat silicious Limestone, filled with masses of coral.

The next succeeding 150 feet are made up of a soft white Limestone, often quite pure and filled with *Orbitoides Mantelli.* The lowermost 60 feet are of impure argillaceous Limestone, which, in disintegrating, yields a black calcareous soil similar to that derived from the rotten Limestone of the Cretaceous. This lower portion of the White Limestone surpasses the others in the variety of its fossils, and most of the forms described in this bulletin from the White Limestone come from this horizon.

2. *The Claiborne.*—The thickness is 140 to 145 feet, and the materials, sands and clays, which are generally calcareous and often glauconitic. Near the top of the subdivision is a bed of glauconitic sand 15 to 17 feet in thickness, filled with shells in perfect state of preservation. The sandy clays forming the lowermost 50 feet are likewise filled with a great variety of shells in good state of preservation. The intervening calcareous clays and calcareous sands are distinguished by the great numbers of shells of *Ostrea sellæformis* which they hold, as well as by the comparative rarity of other forms.

In view of the importance of this subdivision, I give below a more detailed section of the Claiborne Bluff and of the Lisbon Bluff, which is stratigraphically the continuation of the Claiborne Bluff.

This section, which was made jointly by Mr. Aldrich and myself, was first published by Mr. Aldrich in the *American Journal of Science,* October, 1885. In the Geological Report for 1883–'4 substantially the same section is given, but I shall here retain the numbers and letters by which the several beds are distinguished in the section published by Mr. Aldrich, as the references given in this Bulletin are generally to this particular section.

Section of the Bluff at Claiborne.

Drift . ...40 ft.

D. White Limestone bed containing Scutella and casts of shells. Zeuglodon bones found in this bed.—Hale, *American Journal of Science,* 1848, p. 361. .... 45 ft.

C. Scutella bed.   *S. Lyelli* in great numbers ....3 ft.

B. Coarse ferruginous sands, indurated at bottom ....6 ft.

A. Claiborne fossiliferous sand ; source of most of the Claiborne shells ; thin layers of lignite about 10 feet down from top. ....17 ft.

1. Indurated glauconitic sandy ledge......... . ...4 ft.

2. Calcareous clayey strata, becoming sandy in lower part.........18 ft.

3. Indurated sandy ledge..1 ft. 6 in.
4. Calcareous clay, sandy at bottom ....5 ft.
5. Light yellowish-gray calcareous sand, lower part indurated, containing casts of shells...5 ft.
6. Light yellowish-gray calcareous sands, containing *O. sellæformis,* Con.; *Scutella Lyelli,* in fragments; *Scalpellum Eocense,* Meyer; *Pecten Deshayesii,* Lea; *Pecten scintillatus* (?), Con., etc. Shows indurated ledges in places...27 ft.
7. Layer of comminuted oyster shells................................3 ft.
8. Dark blue sandy clay..2 ft.
9. Bluish green clayey sands, very few fossils in upper part, crowded below, a large number of fossils distorted by pressure. *O. sellæformis,* Con.; *Venericardia rotunda,* Lea; *Nucula magnifica,* Lea; *Arca rhomboidella,* Lea; *Anomia n. sp. Amphidesma linosa,* Con., and many other bivalves...10 to 15 ft.
10. Dark bluish-green sand, containing a peculiar small form of *Vener. planicosta,* Lam.; *Turritella Mortoni,* Con.; *Turritella n. sp., Crassatella, sp., Corbula. sp.,* many bivalves...6 ft.

The Section at Lisbon is as follows, and Nos. 1 and 2 are equivalent to Nos. 9 and 10 of the preceding.

Section at Lisbon.

1. Brown, sandy strata, badly weathered, few fossils.................10 ft.
2. Sandy clays, dark brown, badly weathered also, but highly fossiliferous; contain the same fossils as Beds Nos. 9 and 10, at Claiborne, such as the following: *Amphidesma linosa, Arca rhomboidella, Turritella n. sp., V. planicosta,* the peculiar, small variety; *V. rotunda, Lucina compressa, Ancillopsis vetustus, Rostellaria Whitfieldi,* Heilpr, etc., 12 ft.
3. Hard, sandy ledge...8 in.
4. Calcareous, clayey sands, light yellow when wet, nearly white when dry...6 to 8 ft.
5. Coarse-grained, ferruginous sands, fossils numerous................3 ft.
6 and 7. Light yellow sand, with hard ledge on top; lower five feet dark blue when wet...20 ft.
8. Bluish-black clays; first of the *Buhrstone*........................8 ft.

3. *The Buhrstone.*—Minimum thickness 300 feet, the materials almost altogether aluminous and silicious, consisting of aluminous sandstones, claystones and quartzose sandstones, with occasional thin beds of glauconitic sands.

The few fossils which have been obtained from this division are mostly in the form of casts. They do not appear to differ specifically from those of the overlying division.

4. *The Lignitic.*—This is the most massive of the subdivisions of the Tertiary, having a thickness which can hardly be less than 900 feet. It also presents a greater variety in mineral composition, as well as in fossils, than the other divisions. In the most general terms, the Lignitic strata are cross-bedded sands, thin-bedded or laminated sands, laminated clays and clayey sands, and beds of Lignite as well as Lignitic matter, which merely colors the sands and clays.

With these are interbedded, at several horizons, strata containing marine fossils. For the sake of greater convenience and clearness of description, we present the Lignitic in *seven sections*, each of which includes, and is characterized by, one or more beds of marine fossils. These sections are as follows:

1. *The Hatchetigbee Section.*—175 feet in thickness; made up of sandy clays of prevailing brown or purplish color, containing three or four beds of marine fossils in the upper 75 feet, and of somewhat similar purplish-brown, sandy clays, nearly devoid of marine fossils, in the lower 100 feet.

All these brown, sandy clays become much lighter colored upon drying and exposure to the weather.

The following is a more detailed

Section at Hatchetigbee Bluff.

1. Light-colored, aluminous rocks, lowermost of the Buhrstone. ....
....20 to 30 ft.

2. Sandy clays of brown, yellowish and reddish colors interstratified...
....15 to 20 ft.

3. Heavy-bedded, dark-brown clays (similar to 2), but of darker color when dry.. ....10 ft.

4. Yellowish, glauconitic marl (marine shells)............. .....2 to 3 ft.

5. Purplish-brown, sandy clays, in the middle of which is a projecting ledge of dark-colored clays, which are harder, but which break up into small prismatic fragments upon weathering..................................15 ft.

6. Yellowish-gray sands, striped with thin streaks of brown, sandy clay, indurated in places..5 to 6 ft.

7. Bluish-brown, sandy-clay marl, containing many new forms of shells ....5 to 6 ft.

8. Laminated, grayish sands, interstratified with thin beds of brown or black lignitic clay, indurated in places.................................4 ft.

' 9. Heavy-bedded, gray, sandy clays, with streaks of brown colored clays ....8 ft.

10. Reddish, sandy marl, highly fossiliferous, forming concretionary bowlders. Remarkable for the great number of *Venericardia planicosta*, but containing also many other forms, such as *Athleta Tuomeyi, Fusus pagodiformis*, etc............. ....4 to 5 ft.

11. Dark-gray to brown, sandy clays to water's edge. ....15 ft.

2. *The Wood's Bluff or Bashi Section.*—80 to 85 feet in thickness.

The uppermost 30 feet of this section consist of dark-brown clays passing into a green sand, which holds a great variety of finely-preserved marine shells. Below this green-sand marl are gray, sandy clays, with four or five thin beds of lignite within the first 25 feet, succeeded by about 30 feet of cross-bedded sands, with a two-feet seam of lignite at the base.

The details of the most important fossiliferous beds of this section, occurring at the typical locality, are given in the following :

Section at Wood's Bluff.

Orange sand or stratified drift ; surface. ....20 ft.

1. Dark-brown or bluish-black laminated clays, forming the lowermost beds of the Hatchetigbee Section ....10 ft.

2. Dark-bluish, sandy clays, turning red on exposed surface. This bed is highly fossiliferous, containing *Laevibuccinum striatum, Athleta Tuomeyi, Fusus pagodiformis*, and many other forms, some of which seem to be confined to this horizon........ ....3 to 4 ft.

3. Bluish, laminated clay, or sandy clay, much like No. 2 in color and texture, but containing no fossils, or very few, and not turning red on exposed surface ; of variable thickness ....5 to 8 ft.

4. Bluish or greenish sandy clay, somewhat indurated, of decidedly reddish color on surface ; highly fossiliferous ; characterized by *Turritellas* and *Dentalum microstriatum*, but containing also *Ancillaria staminea, Pyrula multangulata, Corbula oniscus, Infundibulum trochiformis*, etc. The lower part of this bed passes gradually into the green-sand marl No. 5, and is the best collecting ground, as the material is less indurated and the shells more easily removed. Thickness ....3 to 4 ft.

5. Green-sand marl, 10 to 12 feet down to the water level. The upper part of the marl is quite soft and friable, but just above the water's edge it becomes indurated, forming rounded bowlderlike masses. This and the preceding might together be considered as one, and called the *lower marl.*

3. *The Bell's Landing Section.*—This is 140 feet in thickness, and includes two important marine beds, and a third quite small and apparently unimportant. These fossiliferous beds are interstratified with yellowish sands in the upper, and rather heavy-bedded sandy clays in the lower, part of the section. The upper marine bed, called the *Bell's Landing Marl,* is about 10 feet in thickness, and has 40 feet of sandy strata above it. The middle bed is called the *Gregg's Landing Marl,* and it lies 20 to 25 feet below the preceding, from which it is separated by dark-gray, sandy clays. This bed is some five or six feet in thickness, and the upper part of it is seen near the water level at Bell's Landing, while the whole bed is well shown at several points higher up the river, notably at Gregg's Landing, Peebles' Landing and Lower Peach Tree. The lowermost of the fossiliferous beds of this section is only about one foot in thickness, and it lies about 50 feet below the Gregg's Landing bed, being separated from it by sandy clays. It is highly glauconitic, but does not contain any great variety of fossils. It has been observed only at Lower Peach Tree. The Bell's Landing Marl is distinguished from all others in Alabama by the great size of the shells which it holds. This bed is seen at the typical locality, Bell's Landing; at Peebles' Landing, at Gregg's Landing, at Lower Peach Tree and at Yellow Bluff on the Alabama, and at Tuscahoma, Turner's Ferry and Barney's Upper Landing on the Tombigbee River.

4. *The Nanafalia and Coal Bluff Section* —The strata of this section are 200 feet in thickness, and consist of about 50 feet of gray, sandy clays at top, which show a great tendency to indurate into tolerably firm rocks, resembling very closely some of the strata of the Buhrstone. Below this, about 80 feet of sandy beds, often strongly glauconitic, characterized throughout by the presence of small oysters, *Gryphœa thirsœ.*

Near the base of this sandy division there is a bed about 20 feet thick, literally packed with these shells.

Below the *Gryphœa thirsœ* beds follow some 70 feet of cross-bedded sands, glauconitic, and apparently devoid of fossils, including, about 10 feet from the base of the section, a bed of lignite which varies in thickness from four to seven feet.

The greatest variety in the shells of this group is seen in the strata which immediately underlie the lowest of the beds containing *Gryphœa thirsœ,* and the following section of the type locality is therefore given :

Section at Nanafalia Landing.

1. Green-sand marl, highly fossiliferous, the main form being *Gryphæa thirsæ*, with a few other shells, such as *Turritella Mortoni*, a *Flabellum*, etc. This is the lowermost of the Gryphæa beds, and is here about 20 feet thick.

2. Dark-blue, almost black, laminated clay, devoid of fossils, but passing below gradually into a bluish marl. ....3 to 4 ft.

3. Bluish, green-sand marl, containing a few shells in the upper three or four feet, but becoming much more fossiliferous below. This bed contains a great variety of beautifully preserved and easily detached fossils, which can, however, be seen only at a low stage of water. Thickness . ...8 to 10 ft.

5. *The Naheola and Matthews' Landing Section.*—It is difficult to give the thickness of the strata of this section, since it varies on the two rivers. We have placed it at 130 to 150 feet. The strata are gray, sandy clays in the main, alternating with cross-bedded sands. The beds of dark, sandy clay containing the marine shells lie at the base of the section; but the Naheola bed on the Tombigbee, while occupying a similar stratigraphical position to the Matthews' Landing bed on the Alabama, differs from it in the character of the materials as well as in the fossils. The two have not been seen so near together as to enable us to determine their relative position.

The following details are given of the

Section at Matthews' Landing.

1. Bluish-black, micaceous, clayey sand, with finely preserved fossils. Very dark when wet, but becoming grayish blue on drying. Crumbles upon exposed slopes, liberating the fossils, which lie in the crumbs thus produced. This is capped by an indurated, sandy, concretionary ledge. Thickness of the fossiliferous strata about.......... ....5 or 6 ft.

2. Gray sands with slightly yellowish cast, showing a tendency to in-durate into lens-shaped bowlders one to two feet thick and three to four feet wide. These sands are also fossiliferous, but much less so than the preceding beds; the fossils are difficult to get out because of the hardness of the material..3 to 4 ft.

3. Bluish, micaceous, clayey sand, much like No. 1, but not holding all of its characteristic fossils. Where this stratum lies exposed to the sun and weather upon flat or nearly horizontal benches, it disintegrates, like No. 1, into crumbs, in which the liberated fossils lie loosely; but

where it forms vertical bluffs, it is firm and compact and resembles black clay. Thickness about.. ....7 or 8 ft.

6. *The Black Bluff Section.*—Here, again, we have difficulty in determining the exact thickness, since, on the Tombigbee, the strata of this section are spread over an extent of surface which would, with uniform dip, correspond to a thickness of over 200 feet, while, on the Alabama, and more particularly inland, in the eastern part of Wilcox County, the thickness is not greater than 35 to 40 feet. Since 80 feet of these beds are seen in superposition at one locality (Black Bluff), we think that the maximum can not be less than 100 feet.

The characteristic strata which compose nearly the whole of this section are black or very dark-brown clays, which are in part fossiliferous. A detailed section would present nothing of importance for our present purpose.

7. *The Midway or Pine Barren Section.*—Thickness 25 feet. The strata are a white, argillaceous limestone, holding a large *nautilus*, which is characteristic of the horizon, 10 feet thick; followed by calcareous sands and a yellowish, crystalline limestone, with *turritellas, carditas* and *corals*—the sands six feet, the limestone eight or nine feet. This section is best seen in eastern Wilcox County, on Pine Barren Creek; but the upper or nautilus-bearing rocks occur at Midway, on the Alabama River, and westward across Marengo County. No exposure of these rocks has been observed on the Tombigbee River, but they will probably be found a short distance below Moscow.

I.

PRELIMINARY REPORT

UPON THE

Tertiary Fossils of Alabama and Mississippi.

BY

T. H. ALDRICH.

Dr. Eugene A. Smith,

    State Geologist, University, Alabama.

    Dear Sir:—The inclosed report consists of a list of the fossils procured by your survey in Alabama, their distribution, with descriptions of part of those believed to be new, also of others collected in Mississippi, and one new species from the Jackson Group of Louisiana. There are added twenty previously described fossils from Alabama and Mississippi, which are all likely to be found in your State.

    At my request Dr. Otto Meyer has added in a separate paper a large number of new species obtained by him partly from the Claiborne Sand, and in the Jackson and Vicksburg Groups in Mississippi.

                          Very Respectfully,

                                T. H. ALDRICH.

Blocton, Bibb County, Alabama, *June* 13, 1886.

PREFACE.

The present condition of the paleontology of the Tertiary Formation is so much confused that it has been found quite impossible to make a complete and accurate report at this time; therefore, simply as a beginning, the writer has prepared the following descriptions and lists.

No attempt has been made to correct the nomenclature; in general, the name under which a species has been described, is used, sometimes that of Conrad, and often that of Lea, without attempting to decide questions of priority.

Again, many species are described without figures, and in others the descriptions are so vague or figures so poor, it has been found out of question to identify from them alone. The writer hopes to correct, as far as possible from the types these uncertainties, and in a final report to bring the subject up to date. In these tables or lists where a species could not be definitely ascertained, no specific name is given; where the identification is at all doubtful, the specific name is followed by a question mark, and where a species is believed to be new, it is so indicated.

At present the collections in hand have not been sufficiently examined to report upon the Claiborne ferruginous sand-bed proper, the Jackson and Vicksburg Groups. These are reserved for the future.

Plates 1 and 2 were published in the Journal of the Cincinnati Society of Natural History, Vol. VIII., July, 1885, but as several changes are necessary, they are inserted here.

The writer is greatly indebted to Prof. Angelo Heilprin, of Philadelphia Academy of Natural Sciences; Prof. R. P. Whitfield, of the American Museum, New York; and Dr. Otto Meyer for identifications and critical remarks. To Dr. C. A. White, of the U. S. Geological Survey for the loan of types; and more especially to Dr. Eugene A. Smith for the opportunity to make the collections from which this paper is prepared.

PART I.

Notes and Descriptions of Species.

In univalves herein described, the term "longitudinal" is understood to be from apex to canal and "transverse' at right angles to this direction. The figures are all either natural size or the true dimensions are indicated with each.

GASTROPODA.

MUREX ANGELUS, Aldr. Pl. 2, fig. 2.

Murex angelus, Aldr. J. C. S. N. H., July, 1885, p. 145. Pl. 2, fig. 2.

MUREX MATTHEWSENSIS, n. sp. Pl. 3, fig. 15.

Shell triangular, whorls probably four; angular, smooth between the varices; varices three, longitudinal, prominent; spines of the body whorl elongated and curved upward, the one at the angle of the aperture nearly closed; body whorl angulated on upper part, the other whorls rounded; aperture ovate, outer lip thick with a foliation at the junction with the body whorl running from the spine to the beak; inner lip smooth with a slight lamina; beak short; canal rather wide.

Locality.—Matthews' Landing, Ala.

Differs from *M. moru'us*, Con., in having no spines between the varices at the shoulder of the body whorl, and has no crenulations on the edge of the outer lip.

MUREX SIMPLEX, n. sp.   Pl. 5, fig. 8.

Shell short, stout; whorls probably five; suture deeply impressed; varices numerous, very large and broadly rounded, terminating above near the suture in sharp points; seven on the body whorl, numerous coarse raised revolving lines cover the whorls; aperture small, elliptical, terminating anteriorly in a nearly closed canal ; outer lip thickened and crenate within ; three folds appear upon the columella.

Locality.—Bryan's Ferry, Miss.; Vicksburg Group.

This type specimen has retained its former apertures, which give to the shell a broad termination and a false umbilicus.

TROPHON GRACILIS, n. sp.   Pl. 5, fig. 6.

Shell acuminate, whorls, ten, rounded ; spine high, with three embryonic whorls, the first two smooth, the next showing longitudinal varices, the balance with numerous (in the type nine) strong varices, which are, when perfect, thin, fringing and sigmoid ; six or more revolving lines cut the edge of the varices into an equal number of crenulations, these revolving lines being strongest at their intersecting points ; aperture ovate, terminating in a narrow canal which turns strongly to the left and slightly upward ; outer lip sharp, thickened and crenulated within ; three slight protuberances on the anterior part of the columella near the canal.

Locality.—Lower Bed, Wood's Bluff, Ala.

Only two specimens found. This species is more acuminate and has a shorter canal than is usual in living forms.

TROPHON CAUDATOIDES, n. sp.   Pl. 6, fig. 4.

Shell with numerous varices; whorls seven, uppermost two smooth, the others angulated by the variceal nodes and crossed by a few revolving lines, which are rather coarse and somewhat alternate; one to four thickened lamelliform varices on different specimens at irregular distances apart. Aperture ovate ; outer lip expanded and crenate within ; columella smooth, twisted below; canal rather abruptly turned to the left.

Locality.—Hatchetigbee Bluff, Ala.

PSEUDOLIVA UNICARINATA, n. sp.   Pl. 5, fig. 17.

Shell broadly ovate, whorls seven, sutural line wavy ; spine sharp, the upper part rising suddenly from the flattened body whorl; nucleus composed of three smooth embryonic whorls. Body whorl shouldered, bearing large longitudinal ribs pointed at the angle, extending nearly to the sulcus below, above reaching to the suture, where they abruptly turn to the left and form rather deep pits betwen them. The carina of the body whorl has a raised line connecting the pointed part of the tuberculations; all the whorls below the nucleus show the tuberculations; sulcus rather deep. No umbilicus; entire surface covered with very fine revolving lines.

Locality.—Matthews' Landing, Ala.

Nearest to *Pseudoliva tuberculifera* Con.; but that species has strong revolving lines; its tubercles are lower down on the body whorl, not sharp; the sutural part is entirely different, and the shell is besides much more fusiform.

The pits above mentioned are very small on young specimens, and are not shown in the figure.

PSEUDOLIVA SCALINA, Heilpr.   Pl. 6, fig. 10.

Pseudoliva scalina, Heilpr.   P. A. N. S., p. 371.   Pl. 20, fig. 12, 1880.

The figures given by Prof. Heilprin do not give a clear idea of this species. Our figure is from the type, and has been redrawn by Dr. Otto Meyer. The original, now in the University Cabinet at Tuscaloosa, Ala., is from Wood's Bluff, Ala., and is extremely rare there.

At Bell's Landing, on the Alabama River, it occurs of gigantic dimensions. The largest specimen in the University Cabinet is $4\frac{8}{16}$ inches long, and $2\frac{7}{16}$ inches wide. The transverse lines are dim from wear, the tuberculations are strongly curved to the right, the plaits on top rise up to the suture of the succeeding whorl abruptly from the shoulder and are bent to the left.

Very close to *Pseudoliva robusta* Briart et Cornet, from the Calcaire Grossier de Mons, France.

TRITON (SIMPULUM) CONRADIANUS, Aldr.   Pl. 2, fig. 8.

Triton (Simpulum) Conradianus, Aldr.   J. C. S. N. H., July, 1885, p. 148.   Pl. 2, fig. 8.

Very common in Eastern Mississippi, in the Red Bluff Strata.

RANELLA (ARGOBUCCINUM) TUOMEYI, n. sp.   Pl. 3, fig. 3.

Shell oblong-ovate, canal strongly recurved, bent upward; whorls seven; spire elevated, pointed, the first two whorls smooth, the others cancellated, the longitudinal lines forming tubercles at intersections ; tubercles sharp, transverse, strongly developed on the periphery of the body whorl and next one above, generally three large ones on the body whorl between the varices; transverse striæ numerous, composed of coarse lines, having three finer ones between, and others between these ; line of growth fine ; varices strong, pitted on the back side ; suture impressed, slightly shouldered ; aperture ovate ; outer lip with a strong varix, nine tubercles within, canal nearly as long as the aperture.

Locality.—Lower bed, Woods' Bluff, Ala.

Young shells show more tubercles between the varices than the type. Named in honor of the late Prof. Michael Tuomey.

FUSUS PEARLENSIS, Aldr. Pl. 1, fig. 17 a, 17 b.

Fusus Pearlensis, Aldr. J. C. S. N. H., July, 1885.

Fusus Boettgeri, Meyer. *Am. Jour. of Science*, June, 1885. (Name preoccupied.)

This species has been found to be quite variable.

FUSUS MEYERI, Aldr., n. sp. Pl. 3, fig. 12.

Shell elongate fusiform ; spire slender, acute ; whorls fourteen ; surface of the spire and body whorl with seven longitudinal broadly rounded folds, which are spirally arranged, crossed by raised rounded striæ, generally seven in number, rather distant, the central one making a sharp carina on the center of each whorl, with erect longitudinal tubercules at the intersections; spaces between striæ showing only lines of growth ; canal very long, spirally striated with alternate raised lines; lines of growth very numerous and almost obsolete ; mouth small, oblong-ovate ; outer lip incurved, smooth.

Locality.—Lower bed, Woods' Bluff, Ala.; Matthews' Landing, Ala.

The figured type retains four embryonic whorls ; three are smooth, the fourth longitudinally striate. The Matthews' Landing form is smaller and even more beautiful. One specimen has the ribs obsolete on the upper half of the whorls and the periphery armed with erect longitudinal spines, giving the shell a very strongly carinate form. *Fusus Mississippiensis*, Con., from Vicksburg, resembles this form, but this is more carinate, with longer canal and spire ; the latter species has many more revolving lines, and the outer lip is striate internally. Named in honor of Dr. Otto Meyer.

FUSUS TOMBIGBEENSIS, n. sp.　Pl. 5, fig. 7.

Shell fusiform; whorls nine, rounded; spire high, apex composed of three embryonic whorls—first two smooth, the next longitudinally ribbed, balance with about ten rounded ribs, somewhat alternating at the suture, strongest on the center of each whorl; whorls crossed by raised rounded lines, alternate in size and nearly equidistant; aperture oblong-ovate; outer lip crenate within; canal short, straight, rather open.

Locality.—Woods' Bluff, Ala.; lower bed.

The type specimen has unfortunately lost its canal, but a younger specimen supplied the description. Younger specimens have the whorls more carinated than in the figured type.

FUSUS RUGATUS, n. sp.　Pl. 5, fig. 9.

Shell fusiform, spire high, suture linear; whorls carinated, concave and smooth above, rounded below, the periphery of each whorl with numerous tubercles, some of them reaching to a second revolving raised line below; the whorl next above the body whorl showing two tuberculated lines below the carina; body whorl showing four rows of spinous lines, contracted rather abruptly below them; canal covered with distant spiral rows of sharp spines; aperture small, angulated posteriorly, terminating in a long, narrow canal.

Locality.—Gregg's Landing, Ala.

The type specimen is broken, but other specimens likewise broken show a long, narrow canal spirally striated to the end.

FASCIOLARIA JACKSONENSIS, Aldr.　Pl. 2, fig. 12.

Fasciolaria Jacksonensis, Aldr.　J. C. S. N. H., July, 1885.

Dr. O. Meyer considers this species identical with his *Turbinella humilior*, *American Journal of Science*, June, 1885, which he now places in the genus *Latirus*. Whether the remarks in his paper constitute a prior description I leave for others to decide.

FASCIOLARIA PERGRACILIS, n. sp.　Pl. 5, fig. 18.

Shell narrowly fusiform; spire very slender; suture impressed; whorls thirteen—nucleus composed of three smooth ones, the following seven are longitudinally ribbed, balance nearly smooth; two equidistant revolving grooves (the one nearest the suture the largest) border it throughout.

Canal long, spirally striated; outer lip smooth; columella bearing posteriorly three faint oblique plaits far within the aperture.

Locality.—Gregg's Landing, Ala.

Decidely fusiform in aspect.

LEUCOZONIA BIPLICATA, n. sp.   Pl. 5, fig. 15.

Shell broadly fusiform; whorls six—nucleus composed of two smooth whorls, those remaining covered with revolving, somewhat alternating raised lines and strong longitudinal ribs, which give the shell an angular appearance. Outer lip sinuate, thickened internally, finely crenated; columella bearing two erect, strong folds, the posterior one nearly twice as large as the other, below these folds nearly straight.

Aperture terminating in a short, open canal. A small umbilicus nearly covered by callus.

Locality.—Matthews' Landing, Ala.

BULBIFUSUS PLEXUS, n. sp.   Pl. 6, fig. 7.

Shell broadly fusiform; whorls probably six, surface on upper whorls with a few distinct revolving lines, which on the upper part of the body whorl become faint; body whorl swollen above, rapidly narrowing below, where it is spirally striated with alternate raised lines; whorls thickened and constricted at suture and suddenly rounded; suture hidden in the groove thus formed; surface finely marked with lines of growth; aperture oblong-ovate, terminating in a rather broad canal, which turned to the left; outer lip slightly crenate within; columella smooth, no perceptible callus on posterior part.

Locality.—Bell's Landing, Ala.

Resembles somewhat the Jacksonian species of *Clavella,* but its whorls are few, spire low and canal recurved. The apex is worn, but appears to be blunt.

BULBIFUSUS INAURATUS, Con.   Pl. 6, fig. 11.

Fusus inauratus, Con.   Foss. Shells of Tert., p. 29.
Fusus Fittoni, Lea.   Cont. to Geol., p. 150.   Pl. 5, fig. 156.
Bulbifusus inauratus, Con.   Cat., 1865.

Figured to show its similarity to the species described below.

BULBIFUSUS TUOMEYI, n. sp.   Pl. 6, figs. 12, 12 a.

Shell large, bulbiform; whorls seven; spire moderate; body whorl very large and globose, flattened above and slightly concave, contracted below and finely striated ; aperture oblong-ovate ; outer lip smooth within ; columella strongly excavated, canal wide and curved.

Locality.—Bell's Landing, Ala.

This species may only be a strongly marked variety of *B. inauratus*, Con., mentioned above, but the younger specimen (12 a) from Gregg's Landing differs from Conrad's species in having finely revolving lines over its whole surface, whorls convex, closely appressed at the suture, while the other has the whorls concave, the lines obsolete on the central part of body whorl, is shouldered at the suture, which is in a groove, and generally has the first four whorls of the spire with a row of revolving nodes above the suture.

NEPTUNEA ENTEROGRAMMA, Gabb. Pl. 3, fig. 5.

Neptunea enterogramma, Gabb. J. A. N. S., Phila. Vol. IV., 2d ser., 1860, p. 378, pl. 67, fig. 14.

This species described from Texas is not uncommon at Lisbon. Gabb's type was a young shell. The two forms are connected by a series in my possession.

NEPTUNEA CONSTRICTA, n. sp. Pl. 5, fig. 13.

Shell broadly fusiform, whorls eight, spire high ; suture distinct, whorls concave and compressed just below the suture ; covered at this point with fine transverse lines, remaining space smooth. Body whorl large, smooth below the above-mentioned sutural area, narrowed below into a short, stout canal which is marked by fine revolving lines. Aperture about two-thirds the length of shell ; outer lip sharp, thickened and crenulate within ; inner lip concave above, straight below ; callus not heavy.

Locality.—Matthews' Landing, Ala.

This species differs from the previous one mentioned, by its compressed space below the suture and absence of a sutural groove.

FULGUR TRISERIALIS, Whitf. Pl. 1, fig. 23 b.

Pyrula Smithii, Sowb. Aldr. in J. C. S. N. H., July, 1885.
Fulgur triserialis, Whitf. *Am. Jour. Conch.*, 1865, vol. I., p. 260.

On first finding this shell it was supposed to be *Pyrula Smithii*, Sowb., but a comparison since made leads me to consider it distinct. On submitting specimens to Prof. Chas. E. Beecher, at Albany, N. Y., he pronounced them identical with the type now in the cabinet of Prof. Hall.

Pyrula juvenis, Whitfield. Pl. 6, fig. 8.

Pyrula juvenis, Whitf. *Am. Jour. Conch.*, vol. I., 1865, p. 259.

Pyrula multangulata, Heilpr. P. A. N. S., 1880, p. 374, pl. 20, fig. 2.

Hitherto unfigured. Prof. Whitfield has kindly furnished me with a drawing of the type. Common in the lower Tertiary, Bell's Landing, Gregg's Landing, Nanafalia Group, Matthews' Landing and Woods' Bluff. It is subject to considerable variation, and is probably the same species quoted by Prof. Heilprin from Woods' Bluff, Ala., as *Pyrula tricostata*, Desh.

Pyropsis perula, n. sp. Pl. 3. fig. 4.

Shell depressed above, spinous; whorls seven ; spire flat, showing only the upper part of the whorls; suture partially concealed by the spines of the next succeeding whorl overlapping.

Surface covered with strong, transverse, raised, cordlike lines, which are armed with short, erect spines on the canal, becoming obsolete at its base. Spines erect, closed, flaring upward and circling the periphery of the body whorl, the one nearest the aperture being the largest. Aperture ovate; outer lip cut by the transverse lines, smooth within; labium strongly reflected, smooth, suddenly expanding at the anterior end of aperture ; callus becomes thick in old specimens; canal long and narrow.

Locality.—Wood's Bluff, Ala., in lower bed ; also Matthews' Landing, Ala.

The largest specimen, if perfect, would be (2″) two inches broad and over (4″) four inches long.

Pisania? dubia, n. sp. Pl. 3, fig. 13.

Shell fusiform ; whorls rounded, about eight in number; spire acute ; surface covered with equidistant revolving lines, which are broadly rounded, the spaces between smooth.

Lines of growth obsolete ; the embryonic whorls are smooth. Aperture oblong-ovate ; canal moderate ; outer lip thickened and striate within ; inner lip smooth, slightly excavated, thickened and angular at junction with canal.

Locality.—Lower bed, Wood's Bluff.

The absence of a callosity at the posterior end of the aperture makes the generic place doubtful.

BUCCINUM MOHRI, n. sp.   Pl. 3, fig. 16, a.

Shell rather solid ; spire high ; apex obtuse ; whorls seven, rounded ; suture rather shallow ; surface smooth.   Lines of growth coarse, showing on the body whorl.   Outer lip strongly reflected, slightly shouldered at its junction with the body whorl.   Aperture semi-lunate nearly two-thirds the length of the shell, smooth internally, terminating in a short, excised canal.

Locality.—Lisbon, Ala.

This species has some resemblance to *Buccinum stromboides*, Herm., from the Calcaire Grossier, of Grignon, but lacks the striations on the lower part of the body whorl, is less swollen in outline, and has a more strongly reflected outer lip.   Named in honor of Dr. Chas. Mohr, of Mobile, Ala.

COMINELLA HATCHETIGBEENSIS, n. sp.   Pl. 3, fig. 6, a. b.

Shell bucciniform, oblong-ovate ; whorls shouldered, with a depressed groove at the suture.   Spire short, smooth ; apex obtuse.   Body whorl strongly shouldered, contracted below, with numerous revolving lines on the basal portion, which are obsolete on the middle part ; the center is flattened, sometimes concave.   Aperture oblong-ovate ; outer lip sinuous, with a distinct, rounded, semi-circular slit on the upper third below the shoulder, smooth internally ; inner lip smooth, reflected, thickened above and below.

Locality.—Hatchetigbee Bluff.   In two different horizons there.

This peculiar form has such a distinct slit in the outer lip (fig. 6), that it may deserve a subgeneric place, though it resembles the living *Cominella maculata*, Martyn, from New Zealand.   In one specimen the shoulder of the body whorl rises so as to almost hide the succeeding whorl, giving the shell a triangular form.

COMINELLA STRIATA, n. sp.   Pl. 5, fig. 4.

Shell ovate, fusiform ; whorls five to six, with fine, transverse lines, shouldered.   Suture in a depressed groove.   Lines of growth sinuous, giving the shell rather a rough exterior.   Body whorl contracted rapidly from the center toward the base.   Striae coarser on the basal portion ; spire about one-third the length of the shell ; apex blunt ; aperture oblong-ovate, smooth within.   Columella broadly reflected ; canal produced canaliculate at base.

Locality.—Hatchetigbee Bluff, Ala.

Differs from the previous species by being striate, its more produced spire and fusiform shape.

The outer tip is broken away, but the lines of growth indicate a semicircular slit, as in the former species.

PHOS VICKSBURGENSIS, Aldr. Pl. 2, fig. 9.

Buccinum Vicksburgensis, Aldr. J. C. S. N. S., July, 1885.

Changed to the genus *Phos*, Prof. Heilprin having indicated its proper place to me.

NASSA CALLI, n. sp. Pl. 5, fig. 5.

Shell oblong-ovate ; spire elevated ; whorls seven, shouldered, and rounded at suture, with a number of revolving lines just below central part of body whorl ; smooth, with five or six striations on the base. Aperture oblong-ovate ; labrum crenulate within ; a strong fold on the columella at the anterior end of aperture, a number of crenulations above this within the opening.

A thick, well-defined callus spreading over the body whorl, thicker posteriorly.

Locality.—Lisbon, Ala.

Named in honor of my friend Prof. R. E. Call, of the State University, at Columbia, Mo.

TURBINELLA (CARICELLA) RETICULATA, Aldr. Pl. 2, fig. 4, a. b, c.

Turbinella (Caricella) reticulata, Aldr. J. C. S. N. H., July, 1885. Since found in lower strata.

TURBINELLA BACULUS, n. sp. Pl. 6, fig. 2, a.

Shell robust, broadly fusiform ; whorls five—two forming a nucleus, next two cancellated. Body whorl large and globose ; revolving striae alternately coarse and fine, well marked, while the longitudinal ones on it become faint.

Aperture over half the length of the shell ; outer lip smooth ; columella with two nearly equal erect plaits ; callus thin, spreading ; canal short, open and recurved. No umbilicus.

Locality.—Bell's Landing, Ala.

A specimen in the State Collection is over twice as large as the type, but imperfect.

VOLUTA SHOWALTERI, n. sp.   Pl. 3, fig. 14.

Shell oval, oblong; whorls five; spire blunted; surface smooth and shining; lines of growth hardly perceptible; suture moderately impressed.

Body whorl and next succeeding shouldered, bearing thereon numerous small tubercles; above the shoulder they are concave.

Aperture two-thirds the length of the shell; columella four plated—the upper one transverse, the middle ones oblique and larger.   Outer lip smooth.

Locality.—Matthews' Landing, Ala.

This species is a true Volute, and belongs to the section *Vespertilio* Klein—Rare.

Named in honor of Dr. E. R. Showalter, of Point Clear, Ala.

MITRA HATCHETIGBEENSIS, n. sp.   Pl. 6, fig. 3.

Narrow fusiform; whorls about ten; suture impressed; nucleus smooth; upper whorls longitudinally ribbed, carinate at their center and tuberculated; the body whorl slightly concave above, with transverse tubercles; whole surface covered with fine, revolving lines.

Aperture narrow nearly half the length of the shell; outer lip smooth; columella straight, with three nearly equal oblique plaits.

Locality.—Hatchetigbee Bluff, Ala.

ANCILLARIA EXPANSA, n. sp.   Pl. 5, fig. 11.

Shell large, ovate, flattened above; whorls about seven, spire moderate; the youngest whorls encircled by a row of tubercles; suture impressed, covered with a thin callus.   Upper whorls and the upper third of the body whorl flattened, then rather abruptly rounded and regularly diminishing toward the base, where it is bounded by a large basal groove.

Aperture oblong-ovate; outer lip thick, a callosity at the posterior; columella smooth, covered with callus, and somewhat twisted.   No umbilicus; no sulcus visible.

Locality.—Lisbon, Ala.

Faint traces of revolving color lines remain.

EXPLERITOMA, nov. gen.

Shell rounded-elliptical; suture linear; whorls few; a small sulcus on the basal part of the body whorl; base canaliculate.

Aperture either filled in or contracted to half the original size by a thickened, circular peristome.   Umbilicus generally covered by this deposit.

Has a general resemblance to the genus *Pterocheilos*, Moore, by reason of its peculiar aperture.   This deposit or filling in is not accidental, as Dr. O. Meyer has, in his collection, a very young shell of similar form, either the same or an allied species, showing the same peculiarity.   It is from the Claiborne sand-bed.

This young shell has the umbilicus exposed, but the aperture is even more filled in than my type specimen, and the spire is flattened.   This genus would be next to *Ancillaria* in the *Ancillarinæ*.

EXPLERITOMA PRIMA, n. sp.   Pl. 5, fig. 1.

Shell rounded-elliptical ; surface probably smooth ; spire moderate ; whorls, four to six ; a well defined but small sulcus upon the lower half of the body whorl ; base canaliculate, but filled up with a deposit from the aperture.

Aperture elliptical, about one-half the indicated size, caused by a thick deposit covering the whole peristome, except a small space on the outer lip.

Locality.—Branch of Satilpa Creek, Ala., in ferruginous sand-bed of Claiborne Group.

Only one specimen found, and that is considerably eroded.   The back of the shell looks like a specimen of *Ancillopsis subglobosa*, Con.

PLEUROTOMA HEILPRINI, Aldr.   Pl. 1, fig. 15.

Pleurotoma Heilprini, Aldr.   J. C. S. N. H., July, 1885.

PLEUROTOMA AMERICANA, Aldr.   Pl. 1, fig. 16.

Triforis Americanus, Aldr.   J. C. S. N. H., July, 1885.

Prof. Heilprin pronounces this species, which I described as a *Triforis*, to be a reversed *Pleurotoma.*   While the lines showing the slit are not very apparent, yet, in deference to his opinion, the change is made.   It is so different from most of the Triforis group, that it is much more likely to be nearer its true place than before.   As the shell is quite common in Jackson, and always reversed, it is probably a good species.

PLEUROTOMA ANITA, Aldr.   Pl. 2, fig. 3.

P. anita, Aldr.   J. C. S. N. H., July, 1885.

PLEUROTOMA LONGIFORMA, Aldr.   Pl. 2, fig. 10, a. b.

P. longiforma, Aldr.   J. C. S. N. H., July, 1885.

PLEUROTOMA PEREXILIS, n. sp.   Pl. 3, fig. 2.

Shell slender, acuminate; whorls eight, convex, surface covered by revolving lines, which in some places have finer ones between, especially on the lower half of the shell. Lines of growth faint, very distinct under a glass; suture bordered below by an impressed line ; slit wide and shallow ; aperture half the length of the shell.

Locality.—Moody's Branch, Jackson, Miss.

This species is exactly the shape and appearance of *Pleuro venusta,* Heilprin, but, on submitting the above specimen to Prof. Heilprin, he stated it to be distinct, as *P. venusta* has its revolving lines arranged in pairs.   The type specimen belonged to the National Museum, but on sending our shell there it could not be found, so a comparison could not be made.

PLEUROTOMA EXILLOIDES, n. sp.   Pl. 3, fig. 9.

Shell slender; spire high; whorls ten, rounded, slightly shouldered below the suture ; a rather strong impressed line just below, with fainter ones still lower.

The first four or five whorls of the apex smooth, the others transversely striate, striations very closely set on the body whorl.   Slit nearly semi-circular; outer lip gently curved ; columella bent ; canal short, curved to the right.

Locality.—Lower bed, Wood's Bluff, Ala.

This species is close to *P. perexilis,* above described, but differs in the breadth of the body whorl, and the slight shouldering of the same.   The slit is larger, and the revolving lines much fainter.

PLEUROTOMA TOMBIGBEENSIS, n. sp.   Pl. 3, fig. 10.

Shell large, rather thick and solid, fusiform; whorls eleven; suture impressed; the spire regularly acuminate ; the upper whorls flattened; lower ones constricted below the suture.

Body whorl constricted about the center, tapering regularly toward the beak ; a large number of fine sinuous revolving lines upon this lower part.

Aperture less than half the length of the shell ; slit small, situated at the lower part of the sutural constriction, its outer edge rising up and rounded ; canal moderate, open, bent a little to the right ; columella with a reflected callus near the the beak.

Locality.—Lower bed, Wood's Bluff, Ala.

On younger specimens the constriction below the suture is obsolete, being replaced by a few faint revolving lines. Resembles *P. longiforma nobis*, but in some respects differs. It is much more fusiform, heavier, the body whorl much more tapering below. The aperture gradually diminishes anteriorly, while in the first-mentioned species, and in P. Gabbi, Con., the canal is long and slender.

PLEUROTOMA TUOMEYI, n. sp. Pl. 3, fig. 11.

Shell narrow, fusiform ; spire high ; whorls twelve, the first three rounded and smooth, next with revolving lines, and the balance strongly carinated with two or more revolving lines some distance apart, with numerous fine ones between, these last alternate. Suture bounded below by a fine raised line ; surface, from first carination to suture, concave, showing exceedingly fine but distinct lines of growth, which spread over the whole shell.

Body whorl with two strong raised carina, the upper one sometimes double, with lesser lines below, alternating with a number of finer ones. Slit narrow, deep and rounded, situated between the raised sutural line and the first carination. Aperture oblong, narrow, terminating in a long, straight canal ; beak slightly turned to the left.

Locality.—Lower bed, Wood's Bluff, Ala.

The largest specimen measured $1\frac{8}{10}''$ in length—a beautifully ornamented shell, quite rare. Prof. Michael Tuomey was the first to visit this locality, where so many of these new species come from.

PLEUROTOMA (ANCISTROSYRINX) COLUMBARIA, n. sp. Pl. 6, fig. 9.

Shell small, fusiform ; whorls ten ; the nucleus smooth, composed of three whorls ; the remaining seven shouldered, and armed with erect spines rising above the edge of the shoulder, giving to the shell a pagoda form ; upper part of whorls smooth, concave, with a faint broken line just below the suture, composed of longitudinal striæ ; lower part of the whorls with nine or ten coarse alternate revolving rows of rounded tubercles, which extend over the spines. Slit rather large, semicircular.

Locality.—Dry Creek, Jackson, Miss.

The body whorl is broken away, rendering it impossible to complete the description. Closely resembles *Fusus pagoda*, Lesson, a living Philippine Island shell.   Having submitted this specimen to Prof. Dall, he pronounces it a good species, and belonging to the group *Ancistrosyrinx,* instituted by him, characterized by a sharp nucleus, and having the last whorl keeled.

CONUS (CONORBIS) ALATOIDEUS, Aldr.   Pl. 2, fig. 11.

C. alatoideus, Aldr.   J. C. S. N. H., July, 1885.

STROMBUS SMITHII, Aldr.   Pl. 2, fig. 6.

S. Smithii; Aldr.   J. C. S. N. H., July, 1885.

APHORAIS GRACILIS, n. sp.   Pl. 5, fig. 14.

Shell fusiform, whorls eight, spire high; embryonic whorls smooth, the others to the body whorl with longitudinal ribs which curve to the left into the suture and are crossed by fine revolving lines; body whorl expanded into a broad outer lip, furnished with two digitations, the posterior one the largest, rising in some specimens into a long, sharp point, strongly grooved to the apex.   The outer lip extends up the spire to the top of the body whorl only; the groove in the anterior digitation rather faint; surface of body whorl with two revolving carina marking these grooves and covered with faint revolving lines; aperture small, inner lip with a spreading callus; canal moderate, terminating in a sharp point slightly recurved.

Locality.—Gregg's Landing, Ala.

Quite common; the digitations are subject to considerable variation, the younger forms have only the posterior one.

The embryonic apex is missing in every specimen.   One example has made a double outer lip; the posterior digitations lying over each other and distinct; the anterior ones coalescent.

CYPRÆA SPHEROIDES, Con.   J. A. N. S., 2d ser., p. 113.   Pl. 11, fig. 6.

A young specimen of this Vicksburg form is in my collection from the Claiborne Sand.

CYPRÆA LINTEA, Con.   Pl. 5, fig. 2.

The original figure does not show any fine lines, and the description says with four approximate equal impressed lines.   I therefore figured this form as new · but on examining the type at Philadelphia Academy, it proved to be like the species figured. There is some confusion here which needs further investigation.

CYPRÆA SMITHII, n. sp.   Pl. 5, fig. 3.

Shell oblong-ovate, rather flat, surface smooth ; labrum crenulate within, smooth on the base and flattened, reflected somewhat and raised above base of shell ; aperture slightly crenulate within, expanded below. *Locality.*—Gregg's Landing, Ala.

Named in honor of Dr. Eugene A. Smith, of Tuscaloosa, Ala.

CASSIS (SEMICASSIS) SHUBUTENSIS, Aldr.   Pl. 2, figs. 5a, b.

Cassis (Semicassis) shubutensis, Aldr.   J. C. S. N. H., July, 1885.

CASSIDARIA BREVIDENTATA, Aldr.   Pl. 1, fig. 20.

Cassidaria brevidentata, Aldr.   J. C. S. N. H., July, 1885.
Cassidaria carinata, Lam. (Aldr.)   J. C. S. N. H., July, 1885.

In the article published in 1885, the first-mentioned form was given a new name on account of a strong varix, and the second form given Lamarck's name.   Upon comparison with the French shell, I have found differences enough to place our shell in another species, and unite it to *C. brevidentata, Nobis,* although it has no varix and is much more ponderous.

CASSIDARIA DUBIA, Aldr.   Pl. 1, fig. 21.

Cassidaria dubia, Aldr.   J. C. S. N. H., July, 1885.

NATICA RECURVA, n. sp.   Pl. 5, fig. 10.

Shell large, globose, smooth, whorls six, spire low ; suture channeled, that part of the whorls within this groove concave, rising to a shoulder. Body whorl very large, flattened on upper part, abruptly rounded below the umbilicus ; aperture semilunar, rounded anteriorly and narrowed at the posterior part ; callus thick, spreading over the body whorl and partially covering the umbilicus.   Umbilicus large, deep, striated within, a thickened callus or rib proceeding from the lower edge of the outer lip, and rounding into the umbilicus.
Locality.—Lisbon, Ala.

The type shows on the body whorl traces of numerous revolving color lines.

CAPULUS COMPLECTUS, n. sp.   Pl. 6, figs. 1, 1 a.

Shell flattened, irregular in form; whorls three, the first two forming a spiral on the body whorl; surface finely striated with concentric lines. Interior edged ; a large horseshoe muscular impression, very slightly impressed, shows the generic position.

Localities.—Wood's Bluff Group, Hatchitigbee, and Lisbon, Ala.

Found occupying the cavities of other shells. Young specimens might readily be mistaken for *Velutina;* older forms resemble an operculum. In young shells the nucleus is about central; in older forms on the right side ; looks like *Pileopsis squamæformis,* Lam., from the Paris basin.

No. 1 is from Wood's Bluff ; 1 a. from Hatchetigbee.

SCALARIA WHITFIELDI, Aldr.   Pl. 1, fig. 18.

Scalaria Whitfieldi, Aldr.   J. C. S. N. H., July, 1885.

SCALARIA OCTOLINEATA, Con.   Pl. 1, fig. 22.

See the *Jour. Cin. Soc. Nat. Hist.* for notes upon this species.

TURRITELLA BELLIFERA, Aldr.   Pl. 1, fig. 13.

Prof. Heilprin considers this form as a variety of *T. humerosa,* Con., but I prefer to let it remain pending a careful review of these difficult forms.

CERITHIUM LANGDONI, Aldr.   Pl. 1, fig. 14.

Cerithium Langdoni, Aldr.   J. C. S. N. H., July, 1885.

CERITHIUM TOMBIGBEENSE, n. sp.   Pl. 3, fig. 7.

Shell elongated, whorls probably twelve; suture distinct, situated in a depressed space ; surface of the uppermost whorls smooth, those following transversely striated—the lower ones, with numerous oblique longitudinal ribs, rising into tubercles upon the center of the body whorl and the two next above ; a wavy line just below the suture, making a slight shoulder to the whorls. Lines of growth sigmoid, fine and numerous upon the body whorl, obsolete above. Aperture ovate, outer lip reflected below, making a short, open canal.

Locality.—Wood's Bluff, Ala.

Only one specimen found ; the mouth is broken away, rendering it impossible to determine its generic position absolutely. The apex is also missing, but is probably sharp. Suggests the genus *Melania.*

MELANOPSIS CHOCTAVENSIS, n. sp.   Pl. 3, fig. 8.

Shell oblong-ovate; whorls six, shouldered; spire obtuse, the upper part generally plicate; suture impressed distinct. Body whorl constricted below the shoulder, smooth on the main part; fine revolving lines below and often a few on the shoulder, sometimes continuing to the apex just below the suture. Aperture oblong-ovate, canaliculate at base; outer lip smooth, slightly thickened within; columella with a thickened and reflected callus.

Localities.—Hatchetigbee, Butler, Choctaw County, Ala.

Resembles closely the figure of *Melanopsis quadrata*, Con., from the Miocene. It is rather variable in form, very common at the typical locality (Hatchetigbee).

MELANOPSIS ANITA, n. sp.   Pl. 5, fig. 12.

Shell ovate, spire short, surface smooth, body whorl large; aperture large, angular posteriorly, caused by a thick deposit of callus; outer lip sharp, thin; callus reflected below, terminating in a small excision.

Locality.—Gregg's Landing, Ala.

This species differs from *Melanopsis Choctavensis*, above described, by its not having a caniculate aperture below the lower spire, and having a larger aperture.

TROCHUS ALABAMENSIS, n. sp.   Pl. 5, fig. 16.

Shell trochiform, whorls seven, suture deeply impressed; surface marked with four to six strong, revolving lines, with finer ones between, crossed by numerous fine longitudinal striations, which are strongest at the intersecting points; base of the body whorl with finer lines. Umbilicus deep, defined by a sharp carina; aperture rounded; outer lip crenate on its edge, the crenulations running some distance within; inner lip smooth.

Locality.—Matthews' Landing, Ala.

Generally is smaller than the type; not rare.

SCAPHANDER PRIMUS, Aldr.   Pl. 2, fig. 7 a, b.

Scaphander primus, Aldr. J. C. S. N. H., July, 1885.

HAMINEA GRANDIS, n. sp.   Pl. 3, fig. 1.

Shell large, very thin, broadly ovate, summit rounded, with a deep pit marking the position of the spire. Surface covered with narrow, transverse striæ, with rounded spaces between; spaces below the center again subdivided by more shallow lines, closer set as the base is approached, but nearly obsolete at apex; lower extremity obliquely but broadly rounded; aperture rather large, nearly equal in width as far as shown. Inner lip represented by an exceedingly thin lamina, reflected, showing in the type a narrow, oblique umbilicus.

Locality.—Bunker Hill, La.; Jackson Group.

This species is the largest yet described from the Southern Tertiary. The specimen is partly a cast; substance of shell is thinner than ordinary writing-paper. The lines of growth and transverse striæ are well shown on cast.

HAMINEA ALDRICHI, Langdon.

Bulla (*Haminea*) *Aldrichi*, Langdon. *Am. Jour. Sci.*, March, 1886. *Bulla biumbilicata*, Mr. P. A. N. S., Philadelphia, 1884, fig. p. 110.

Dr. Meyer's name is preoccupied by Deshayes' in *Des Animaux sans vertebres dans Le Bassin de Paris*, Atlas, vol. 2, pl. 39, figs. 33, 34 and 35. It is quite likely, however, that this shell may prove to be the true *biumbilicata* of Deshayes.

PELECYPODA.

TEREDO CIRCULA, n. sp.

Consisting of a long, circular, shelly sheath, terminating at the smaller end in two siphons, which are also nearly circular in section. These siphons are partially separated in one specimen, and one is larger than the other; substance of the sheath rather thick; shell smooth at the outer end and annulated and enlarged at the further end. Generally straight or only slightly contorted.

Localities.—Choctaw Bluff, Ala., and Wayne County, Miss., in White Limestone; Vicksburg Group.

Found boring in the limestone in great numbers, standing vertically; the valves not found. Mentioned in previous article as *Aspergillum* tubes, but having since visited Choctaw Bluff and found the terminal tubes, I have ventured to give it a name. No imbricated ledges found in any of the specimens.

PHOLAS ALATOIDEA, n. sp. Pl 4, figs. 9, 9 b, 9 c.

Shell elongate, cylindrical, posterior end concentrically striated, anterior half of the shell crossed with raised radial lines forming imbrications, which grow stronger as we approach this end. Anterior dorsal margin winged, this part showing only the continued concentric lines; umbonial processes large, reflected; anterior ventral margin strongly notched. Internal process broad, spatulate.

Localities.—Gregg's Landing and Bell's Landing, Ala.

This shell was no doubt the one found by Prof. Tuomey and given in his list of species at Bell's Landing under the name of *Pholas Roperiana.* He never described it; the figure (9) nine is erroneously numbered 4' by the lithographer.

MARTESIA ELONGATA, n. sp. Pl. 4, fig. 10.

Shell gaping above, elongated, sub-cylindrical; the anterior and central part of surface marked with concentric ribs, balance of posterior smooth. A strongly impressed groove running from the beaks to the ventral margin, the concentric ribs being more prominent on the sides of this groove.

Beaks strongly recurved toward the anterior, situated close to this end, which is truncate.

Locality.—Yellow Bluff, Ala., Bell's Landing Group.

This species was taken from a piece of lignite by D. W. Langdon, Jr. The shell has the markings of the dorsal accessory plate.

SOLEN LISBONENSIS, n. sp. Pl. 4, fig. 4.

Shell linear, nearly straight; posterior sub-truncate; anterior obliquely rounded with a depressed space behind running to the beaks; lines of growth prominent, bending at right angles along a line running obliquely from the beaks to the junction of the posterior and ventral margins. Anterior widely gaping.

Locality.—Lisbon, Ala., just above the Buhrstone.

MACHA VICKSBURGENSIS, Aldr. Pl. 2, fig. 1.

Solecurtus Vicksburgensis, Aldr. J. C. S. N. H., July, 1885.

PANOPÆA PORRECTOIDES, n. sp. Pl. 4, fig. 3.

Shell thin, oblong, ventricose; surface marked by lines of growth; anterior side expanded, posterior produced. Ventral margin nearly straight. Beaks incurved, pointed, wrinkled by lines of growth. The posterior end widely gaping, and has a wedge-shaped appearance. Tooth strong, blunt below, sharp above; hinge line short.

Locality.—Baker's Bluff, Ala., Ferr. Sand Bed.

The differences between this form and the Miocene *P. porrecta*, Con., are very slight. Left valve figured.

NEÆRA PRIMA, n. sp. Pl. 6, fig. 14.

Shell rotund, covered with rounded, close-set concentric striæ; posterior with four radiating ribs; anterior part smooth. Rostrum moderate, narrow, rounded, with a concave, almost triangular space between it and the body of the shell, the end curving upward; ventral margin hollowed out; dorsal margin rising slightly above the hinge line. Cartilage pit, minute, bent inward in left valve, which also has the posterior muscular impression strongly defined by a rib on the inner side running to the dorsal margin; hinge line nearly straight.

Locality.—Wood's Bluff, Ala. (lower bed).

NEÆRA ALTERNATA, n. sp. Pl. 6, fig. 15.

Shell small, moderately rotund, covered with very fine concentric lines; rostrum short, obtuse, rounded on top and angulated on dorsal margin; lines of growth becoming lamellar on its surface. Hinge line and dorsal line of rostrum straight. Cartilage pit minute, narrow, oblique and close under the beaks.

Locality.—Lisbon, Ala.

Distinguished from the preceding species by its lack of radiating ribs, smoother surface, and is much less rotund. Has a strong resemblance to the recent *Neæra lamellosa*, Sars.

PHOLADOMYA CLAIBORNENSIS, n. sp. Pl. 4, fig. 5.

Shell exceedingly thin, pearly, concentrically ribbed, crossed by radial lines, which are strong on the umbones; beak high, very prominent close to the anterior end of shell, anterior truncated abruptly; posterior produced, flattened on the upper part; lunule long and narrow, smooth; hinge raised, prominent.

Locality.—Lisbon, and beds at base of Claiborne Bluff, Ala.

This species is close to *P. Marylandica*, Con., but appears to differ from that species in height of beaks and their more anterior position, besides being abruptly truncated and posterior flattened.

ASTARTE CONRADI, Dana, sp.   Pl. 4, fig. 7.

Shell thin, oblong-ovate, concentrically wrinkled with broad, rounded plaits, concavities between nearly smooth.   Covering the whole surface are fine, concentric lines, stronger on top of the plications; beaks flattened, blunt, and turned toward the anterior; lunule heart-shaped, separated by an indistinct ridge; hinge line nearly straight.

Localities.—Lisbon, Coffeeville, and beds at base of Claiborne Bluff, Ala.

The dentition removes this species to *Lutraria*.   Prof. Marsh, at the request of Prof. J. D. Dana, has kindly allowed me to examine the type specimens.   The one figured by Prof. Dana has been considerably distorted by pressure, and therefore misled Prof. Heilprin, who considered it a young *Crassatella alta*, Con.   I figure the normal form.   Ranges from the calcareous sand-bed down to the Buhrstone.   The largest specimen found measures nearly two inches in length.   Very close to *L. lapidosa*, Con.

CYTHEREA HATCHETIGBEENSIS, n. sp.   Pl. 4, fig. 1.

Shell rather thick, inflated, subrotund, transversely striate—the different periods of growth marked by a dropping down of the concentric lamina, giving a ridge-like appearance; umbones swollen; beaks elevated; lateral tooth in left valve transverse, conical and strong; ligament short-curved; excavation of the pallial impression angular; margin of the shell entire, thickened in some specimens.

Locality.—Hatchetigbee Bluff, Ala., beneath the Buhrstone.

Prof. A. Heilprin considers this shell as *C. discoidalis*, Con., but that is described as having the inner margin crenulate, while this is smooth.   Conrad's species has never been figured.

CARDIUM HATCHETIGBEENSE, n. sp.   Pl. 4, figs. 12, 12a, 12b.

Shell large, subquadrate, ventricose with about thirty-two ribs; ribs flattened oval, with the scars of spines visible along their center, a few coarse, flattened, triangular spines remaining on the posterior and anterior; largest on the posterior, which is subtruncate, the largest rib forming the angle; marginal serrature largest at the angle.   The flattened spaces between the ribs are equal in width to the ribs and nearly smooth, showing faintly the lines of growth.   Cardinal teeth in left valve double—the lower one the largest, very strong and erect.

Locality.—Hatchetigbee Bluff, Ala.

CARDIUM TUOMEYI, n. sp.   Pl. 4, figs. 13, 13a.

Shell ovate, thick through the umbones ; ribs numerous, about forty-four in number, flattened above and indented with scars of spines; spaces between the ribs much smaller than the ribs themselves ; a few small, scattered spines near the beaks, which are central and raised ; muscular scars strongly marked.

Locality.—Nanafalia, Ala.

Differs from the previous species described by its more numerous ribs, more rounded form, smaller spines and thicker shell. The figure 13a shows the teeth of the left valve.

CARDIUM EUFAULENSE, Con.   Pl. 6, fig. 5.

A cretaceous fossil (or described as such) received from Dr. William Spillman, of Mississippi.   Probably an error in locality.

VERTICORDIA EOCENSIS, Langdon.   Pl. 6, fig. 13.
V. eocensis, Langdon.   *Am. Jour. Sci.*, March, 1886.

Figured at the request of Mr. Langdon.

NUCULA MONROENSIS, n. sp.   Pl. 4, fig. 2.

Shell ovate-elliptical, subangular behind ; surface covered with raised concentric ribs, sharp on their upper edge and slope downward toward the ventral margin ; beaks recurved ; lunule faintly marked, but large ; cavity of the shell rather deep ; margin crenulated.

Locality.—Calcareous sand-bed lower part of Claiborne Section, Monroe County, Ala.

CUCULLÆA TRANSVERSA, Rogers.   Pl. 4, fig. 11, 11a.

The figure is from a specimen found at Gregg's Landing, Ala. Very large individuals two and one-half inches in length occur. Prof. Heilprin has placed the genus *Cucullæa* in *Arca* as a subgenus, and changes the name of this species to *C. Rogersi*, because the name *transversa* is preoccupied in *Arca ;* but as both Tryon and Zittel consider the genus a good one, the old name ought to stand.

PECTEN (PLEURONECTIA) ALABAMENSIS, n. sp.   Pl. 4, fig. 8.

Shell small, suborbicular; upper valve covered with equidistant concentric lines which run over upon the anterior ear ; a few raised radial lines upon the center and anterior side ; ears small ; right valve nearly smooth ; within both valves eight raised prominent rounded ribs, becoming obsolete as they approach the beak.

Locality.—Matthews' Landing, Ala.

Seems to unite *Pleuronectia* and *Pecten.* One specimen shows concentric striæ and ribs in the younger part of the shell, these becoming obsolete toward the ventral margin. The enlarged figure gives a view of the interior.

ANOMIA EPHIPPIOIDES, GABB., VAR., LISBONENSIS, new var. Pl. 4, fig. 6.

Shell thin, pearly, suborbicular; upper valve smooth, slightly wrinkled on the umbo; lines of growth distant, hinge line doubly sigmoid, the extremities winged. Muscular scars indistinct.

Locality.—Lisbon and beds at base of Claiborne Bluff.

The type is externally marked with broad radiating bands of color. On comparison with A. ephippioides, Gabb., it appears much larger and more transverse; that species is not smooth externally, and is often plicate. It occupies the same horizon however.

OSTREA JOHNSONI, n. sp. Pl. 6, fig. 6.

Shell large, thick ; both valves convex ; beaks pointed in lower valve ; breadth of shell greater than length ; both valves strongly plicate, generally with six folds, the plications becoming very deep with age, the concave part between the two basal folds, running out into a long curved tongue ; surfaces strongly laminated, laminæ terminating at edge of shell ; attachment scar not visible ; ligamental area with a rather deep transversely striated furrow in the lower valve, shallower and broad in the upper; muscular scar, large nearest the base and posterior margin; curved, spatulate, nearly straight on upper side.

Locality.—Monroe County Calc. Sand-bed, Claiborne and Lisbon, Ala., also Newton, Miss.

Named in honor of Lawrence C. Johnson, of the U. S. Geological Survey. This form seems to be confined to the Lower Claibornian above the Buhrstone.

PART II.

Geological Distribution and Localities of Species.

When the species mentioned are in my cabinet, or that of the State University or quoted from some previous author, appropriate signs are used as follows :

My cabinet, .	*
State collection, .	‡
Prof. A. Heilprin, .	†
T. A. Conrad, .	‖
S. G. Morton, .	§
Dr. O. Meyer, .	¶

Nearly all the species, however, are either in the State University cabinet or my own. For the stratigraphical succession of the different groups, the reader is referred to Dr. E. A. Smith's explanatory article accompanying this report.

TABLE I.

WHITE LIMESTONE GROUP.

This formation needs more careful examination, especially in its upper beds. The western part of Alabama shows the beginning of the change that takes place in Mississippi. The most peculiar "find" in connection with this group is the presence of *Ven. planicosta*, Lam. One large valve was found on the top of a high Limestone hill, five miles north of Claiborne, by the writer.

Many of the shells exist only as casts, rendering the absolute determination almost impossible.

Following the names, column one shows those found in the vicinity of Claiborne.

2. those found at a locality in Clark County, Ala., S. 35, T. 9, R. 6, called "The Rocks."

3. Choctaw Bluff, Clarke County, Ala., the most southerly extension of the Tertiary on the Alabama River.

4. St. Stephens' Bluff, on Tombigbee River. The column on the left shows whether the species is in my cabinet or not.

My Cabinet	Name.	Vicinity of Claiborne	"The Rocks," Clark Cr., Ala.	Choctaw Bluff, Ala.	St. Stephens	Remarks.
*	Nautilus Alabamensis, Mor.	*				
*	Cypræa fenestralis, Con		*			
*	Scalaria, sp		*			
	Conus sauridens? Con. (cast)		*			
*	Turritella, sp. (impression)	*				
*	Pecten perplanus, Mor	*	*		*	
*	" Poulsoni, Mor	*		*	*	
	" anatipes, Mor	*				
*	Ostrea Mortonii Gabb (= panda, Mor. pars.)	*				
*	" cretacea, Mor		*			
*	" Georgiana, Con		*			
*	" sp					
*	" Tuomeyi? Con					
*	Gryphæa vomer, Mor		*	*	*	
*	Spondylus dumosus, Mor		*	*	*	
*	Terebratulina lachryma, Mor.		*	*		
	Modiola cretacea, Con					
*	Crassatella, sp. (cast)			*		
*	Cytherea, sp. (cast)	*		*		
*	Leda, sp					
*	Lithodomus, sp. (casts)	*				
*	Pectunculus, sp. (cast)	*				
*	Teredo circula, Aldr.				*	Miss.
*	Venericardia planicosta, Lam	*				
*	Orbitoides Mantelli, Mor	*	*	*	*	
*	Scutella Rogersi, Mor	*				
*	" Lyelli, Con	*		*		
*	" crustoloides, Mor	*				
	Serpula, sp		*			
*	Coral, massive					
*	Echinus infulatus? Mor	*				
*	Cidaris, sp	*				
*	Sword of Cœlorhynchus					
*	Carcharodon augustidens, Ag			*		
	Zeuglodon cetoides					

The massive coral is also found at Central Salt Works, Ala., where it comprises the greater part of a hill over one hundred feet high.

TABLE II.

Fossils from the "Ferruginous Sand-bed" (Claiborne sand) at base of Baker's Bluff, one mile above St. Stephen's Bluff, on Tombigbee River. As the stratigraphical position of this locality has been disputed, this table is here given :

Pseudoliva vetusta, Con.
Ancillopsis subglobosa, Con.
Trochita trochiformis, Lam.
Melongena alveata, Con.
Rostellaria (Calyptrophorus) alveata, Con.
Oliva Alabamensis, Con.
Ancillaria staminea, Con.
Caiicima galba, Con.
Fasciolaria pyruloides, Con.
Voluta petrosa, Con.
Caricella doliata, Con.
Tornatella bella, Con.
Crepidula lirata, Con.
Hipponyx pygmæa, Lea.
Bulla biumbilicata, Mr.
Solarium elegans, Lea.
Dentalium thalloides, Con.
Mesalia obruta, Con.
" striata, Lea.
Cadulus, sp.
Pleurotoma Desnoyersi? Lea.
" elaborata? Lea.
Natica limula, Con.
" ætites, Con.
Cerithium, sp.?
Mitra fusoides, Lea.
Actæon pygmæus, Lea.

Pecten Deshayesii, Lea.

Venericardia alticosta, Con. (transversa,
" rotunda, Lea. Lea.)
" Sillimani, Lea.
Avicula limula, Con.
Cytherea Poulsoni, Con.
" perovata, Con.
" æquorea, Con.
Crassatella protexta, Con.
" alta, Con.
Astarte Nicklinsii, Lea.
" sulcata, Lea.
Leda protexta, Con.
" plicata, Lea.
Corbula oniscus, Con. var. carinata,
" nasuta, Con. [Con.
Arca rhomboidella, Lea.
Pectunculus stamineus, Con.
Nucula magnifica, Con.
Egeria rotunda, Lea.
Lucina papyracea, Lea.
" subvexa, Con.
Panopæa porrectoides, Aldr.

Endopachys maclurii, Lea.
Flabellum, sp.
Discoflustrellaria Bouei, Lea.
Capularia discoidea, Lea.
Heteractis Duclosii, Lea.
Lunulites interstitia, Lea.

The Claiborne Sand-bed also occurs bearing the identical species found at the typical locality at

Stave Creek, Ala., S. 18, T. 7, R. 2, E.
Satilpa Creek, Clark County.
Allen's Creek, Clark County.
At Mr. White's in Monroe County.
These localities will be reviewed in a final report.

TABLE III.

MIDDLE AND LOWER CLAIBORNIAN.

This list includes all the strata between the Calcareous Sand-bed and the next group beneath (Hatchetigbee). There are three horizons in the Claibornian, which are rich in fossils and present features peculiar to each.

1st. The famous Claiborne "Furruginous Sand."

2d. The bed near the base of the Claiborne Bluff called the "Calcareous Sand-bed."

3d. The Lisbon strata down to the Buhrstone.

Excepting the Buhrstone, all the localities given below belong to the second and third horizons—as follows :

Horizon of the Calcareous Sand-bed.

1. Calc. sand-bed proper at Claiborne.
2. T. A. Rumbley's, Monroe County, Ala.
3. Coffeeville Landing, Ala.
4. Near Barrytown, Ala.

Lisbon Horizon.

5. No. 9, Claiborne section (*Am. Jour. Sci.*, October, 1885, and above in Dr. Smith's Summary).
6. No. 10, Claiborne section.
7. Lisbon, in general.
8. Lowest beds at Lisbon.
9. Buhrstone.

(All the species are in my cabinet.)

Name.	1 Calc. Sand.	2 Monroe Co.	3 Coffeeville.	4 Near Barrytown.	5 No. 9, Claiborne.	6 No. 10, Claiborne.	7 Lisbon.	× Lowest Lisbon.	9 Buhrstone.
Ancillaria altile, Con........................								*	
" subglobosa, Con......................					*			*	
" staminea, Con	*							*	
" expansa, Aldr								*	
Agaronia punctulifera, Gabb...........								*	
Rostellaria Whitfieldi, Heilpr..........								*	
" velata, Con.............................								:	
Siliquaria vitis, Con........................							*	*	
Monoptygma Leai, Whitf....								*	
Mitra fusoides, Lea.								*	*
" dumosa, Con., var........................								*	

Name.	1 Caleb Sand.	2 Monroe Co.	3 Coffeeville.	4 Near Barrytown.	5 No. 9, Claiborne.	6 No. 10, Claiborne.	7 Lisbon.	8 Lowest Lisbon.	9 Buhrstone.
Mitra Haleanus, Whitf							*		
" bolaris, Con		*					*		*
" biconica, Whitf							*		
Murex engonatus, Con							*		
Terebra gracilis, Con							*		
" multiplicata, H. C. Lea							*		
" divisura, Con., var							*		
" plicifera? Heilpr							*		
Marginella incurva, Lea							*	*	
Dentalium arciformis? Con		*					*	*	
" thalloides, Con							*		
" multistriatum, Heilpr		*					*		
Hipponyx prgmæus, Lea								*	
Tuba antiquata, Con								*	*
Odontopolys compsorhytis, Gabb					*	*	*		
Trochita trochiformis, Lam		*			*		*		
Natica limula, Con							*		
" gibbosa, Lea					*	*	*		
" minor, Lea							*		
" mamma, Lea							*		
" Newtonensis, M. & A							*		
" sp						*		*	
" Lisbonensis, Aldr		*				*	*		*
Sigaretus bilix, Con							*		
" declivis, Con						*			
" (Sigaticus) Bœttgeri, M. & A							*	*	
Crepidula lirata, Con		*	*				*		
Pseudoliva vetusta, Con		*			*	*	*		
Turbinella pyruloides, Con							*		
Conus sauridens, Con							*		
Voluta petrosa, Con						*	*		
" Sayana, Con., var					*		*		
Turritella Mortoni, Con		*				*	*		
" humerosa? Con							*		
" n. sp						*	*		
" eurynome? Whitf						*	*		
" nasuta, Gabb							*		
" carinata, Lea		*			*	*	*		
Mesalia obruta, Con		*			*	*	*		
" vetusta, Con							*		
Fusus raphanoides, Con							*		
" crebissimus? Lea							*		
" sp							*		
" limulus, Con							*		
" pagodiformis, Heilpr							*		
" Mortoniopsis, Gabb									
" irrasus, Con							*	*	
" trabeatus, Con., var					*		*		
" sp									
Cornulina armigera, Con							*		

Name.	1 Oak Sand.	2 Monroe Co.	3 Coffeeville.	4 Near Barrytown.	5 No. 9, Claiborne.	6 No. 10, Claiborne.	7 Lisbon.	8 Lowest Lisbon.	9 Buhrstone.
Mazzalina pyrula, Con							*		
Cadulus abruptus, M. & A.		*					*		
Buccinum Mohri, Aldr..							*		
Neptunæa enterogramma, Gabb						*	*		
Capulus complectus, Aldr							*		
" n. sp?..							*		
Cylichna galba, Con							*		*
" subradius, Mr									*
" Dekayi, Lea.							*		
Bulla Aldrichi, Langdon	*								
Nassa Calli, Aldr							*		
" cancellata, Lea.		*	*						*
Phos Texana, Gabb.							*		
Cerithioderma, n. sp							*		
Cerithium vinctum, Whitf							*		
Pleurotoma terebralis, Lam							*		
" sp							*		
" sp		*				*	*		
" sp							*		
" sp							*		
" sp							*		
" sp							*		
" sp							*		
" sp							*		
" sp							*		
" sp									*
Cassis Taitii ? Con							*		
Fasciolaria polita, Gabb							*		
Ringicula biplicata ? Lea		*					*		*
Oliva bombylis, Con							*		
" Alabamensis, Con			*						
Pyrula cancellata, Lea							*		*
Actæon elevatus, Lea							*		
" lineatus, Lea							*		
" melanellus, Lea		*					*		
" pygmæus, Lea							*		
Turbo naticoides, Lea		*							
Solarium cancellatum, Lea		*							
" scrobiculatum, Con							*		
" ornatum, Lea							*		
" sp							*		
" elegans, Lea., var							*		
Teinostoma subrotunda, Mr							*		
Delphinula plana, Lea							*		
" dipressa, Lea							*		
Planaria nitens, Lea							*		
Cancellaria alveata, Con							*		
Busycon, n. sp							*		
Scalaria, sp							*		
Odostomia, 2 sp									*

Name.	1 Calc. Sand.	2 Monroe Co.	3 Coffeeville	4 Near Barrytown	5 No. 9, Claiborne	6 No. 10, Claiborne	7 Lisbon	8 Lowest Lisbon	9 Buhrstone
Nautilus, sp.									*
Belosepia ungula, Gabb.		*					*		
Lucina compressa, Lea		*					*		
" Claibornensis, Con		*		*			*		
" impressa, Lea							*	*	
" subvexa, Con		*					*		*
" papyracea, Lea		*					*		*
" lunata, Lea		*					*		
" carinifera, Con		*							
Amphidesma linosa, Con					*	*	*		
" tellinula, Con								*	
Crassatella alta, Con							*		
" protexta, Con		*			*		*		
" palmula? Con						*			
Alveinus minutus, Con						*			
Venericardia, densata, Con					*	*	*		
" rotunda, Lea		*					*	*	*
" Sillimani, Lea		*					*		
" parva, Lea	*								*
Avicula limula, Con	*	*			*		*		
Solen Lisbonensis, Aldr		*					*		
Cardium Nicolleti, Con., var		*							
Petunculus staminens, Con		*					*		
Limopsis declivis, Con					*	*	*		
" decisus, Con					*	*	*		
" ellipsis, Lea		*					*	*	*
" cuneus, Con							*		
Trigonocælia ledoides, Mr.							*		
Astarte parva, Lea							*		
" minor, Lea							*		
" sp							*		
" Couradi, Dana	*			*		*	*		
Mactra prætenuis, Con							*		
" parilis, Con							*		
" Grayi, Lea							*		
Corbula oniscus, Con							*	*	
" nasuta, Con		*			*	*	*		
" gibbosa, Lea			*				*		*
" Murchisonii, Lea (var. fossata, M. & A.)		*							
Pecten Dehayesii, Lea	*	*	*				*		*
" scintillatus, Con	*				*		*		
" calvatus? Mor							*		
Arca rhomboidella, Lea			*		*	*	*		
Bysoarca Mississippiensis, Con	*	*							
Leda media, Lea					*		*		
" multilineata, Lea							*		
" n. sp								*	
Neæra alternata, Aldr							*		
Nucula magnifica, Con	*	*			*	*	*		

Name.	1 Calc. Sand.	2 Monroe Co.	3 Coffeeville.	4 Near Barrytown.	5 No. 9, Claiborne.	6 No. 10, Claiborne.	7 Lisbon.	8 Lowest Lisbon.	9 Bahrstone.
Nucula ovula, Lea							●		
" coelata, Con							●		
" pulcherrima, Lea							●		
" media, Lea							●		●
" semen, Lea		●					●		
" Claibornensis, Con					●	●	●		
" Monroensis, Aldr		●							
Cytherea Poulsoni, Con		●			●		●		
" perovata, Con		●			●		●		
" aequorea, Con	●						●		
" minima, Lea							●		
" trigoniata, Lea					●				●
Plicatula filamentosa, Con									●
Kelliella Boettgeri? Mr							●		
Teredo, sp.?									●
Tellina nitens, Lea							●		
" ovalis, Lea					●	●	●		
" scandula, Con							●		
Ostrea Johnsoni, Aldr		●					●		
" sp							●		
" sellaeformis, Con	●	●	●		●	●	●		
" divaricata, Lea	●	●	●			●	●		
Anomia ephippioides var. Lisbonensis, Aldr	●				●		●		
Egeria rotunda, Lea		●			●		●		
" plana, Lea							●		
" sp.?		●							
Dosinia, n. sp					●	●			
Hindsiella, n. sp. (not faba, Mr.)							●		
Grateloupia Moulinsii, Lea							●		
Pholadomya Claibornensis, Aldr					●	●			
Pinna, sp. (very large)		●	●						
Lunulites Duclosii, Lea		●					●		
Flabellum Wailesii? Con			●				●		
Turbinolia Maclurii, Lea	●	●					●		
" pharetra, Lea							●		
Lunulites Bouei, Lea							●		
Platytrochus Stokesii, Lea							●		
Madrepora, 2 sp		●					●		
Cyclosmilia?			●				●		
Serpulorbis, sp.?							●		
Serpula ornata, Lea		●							
Scutella Lyelli, Con	●		●						
Echinus ——, sp							●		
Scalpellum eocense, Mr	●	●							
Mylobates ——		●					●		
Otolithus		●					●	●	
Foraminifera (1 sp.)							●		

TABLE IV.

HATCHETIGBEE GROUP.

Found at Hatchetigbee Bluff on the Tombigbee River beneath the Buhrstone. A large bed of *Ven.* planicosta, Lam., about two feet thick is seen here at low water, and millions of this species could be obtained. The shore line of the river looks like an ocean beach from the abundance of this form.

Especial attention is called to the very remarkable fauna found here.

Natica limula? Con.
" Mississippiensis? Con. (*umbilicated var.*).
" parva, Lea.
Sigaretus (Sigaticus) Clarkeanus, Aldr., mss.
Nassa cancellata, Lea.
Cerithium, n. sp.
Melanopsis Choctawensis, Aldr.
Voluta petrosa? Con.
V. (Athleta) Tuomeyi, Con.
Pseudoliva vetusta, Con.
Ancillaria subglobosa, Con.
Oliva bombylis, Con.
Cominella Hatchetigbeensis, Aldr.
" striata, Aldr.
Cornulina armigera, Con.
Fusus trabeatus, Con.
" pagodiformis, Heilpr.
" subscalarinus, Heilpr.
" subtenuis, Heilpr.
" sp.?
" 2 sp.
Turritella, sp.
" carinata, Lea.
Rostellaria trinodifera, Con.
Pyrula juvenis, Whitf., (multangulata, Heilpr.).
Dentalium, sp.
Capulus complectus, Aldr.
Trochita trochiformis, Lam.
Tornatella bella, Con.
Cadulus, sp.
Mitra dumosa, Con., var.
" Hatchetigbeensis, Aldr.
Pleurotoma, 4 sp.
Trophon caudatoides, Aldr.
Triton, n. sp.
Columbella, sp.
Cancellaria alveata, Con.
Solarium elegans, Lea., var.

Solarium granulatum, Lea.
Tuba antiquata, Con.
Scalaria sessilis? Con.

Cytherea Hatchetigbeensis, Aldr.
" Poulsoni? Con.
" sp.
" Nuttalliopsis, Heilpr.
Cardium Nicolleti, Con., var.
" Hatchetigbeensis, Aldr.
Venericardia planicosta, Lam.
" Sillimani, Lea.
" alticosta, Con.
Spondylus dumosus, Mor., var.
Corbula nasuta, Con.
Pectunculus stamineus, Con.
Ostrea Vicksburgensis? Con.
" cretacea, Mor.
" sellæformis, Con.
Gryphæa vomer, Mor.
Pecten, n. sp.
Leda protexta, Con.
" sp.
Astarte Nicklinsii, Lea.
Arca subprotracta, Heilpr. (protracta, Con.)
Crassatella producta, Con.
Plicatula filamentosa, Con.
Lucina rotunda, Lea.
" subvexa, Con.
" sp.
Mactra prætenuis, Con.
" n. sp.
Tellina papyria, Con.
Nucula plicata, Lea.
Mysia, sp.

Madrepora, sp.
Cyclosmilia, sp.

Otolithus.

TABLE V.

WOOD'S BLUFF GROUP.

This group has even a greater thickness than the Claiborne Sand, and is fully as crowded with fossils, though there is not so great an assemblage of species. The lower part (twelve to fifteen feet thick) is a mass of *Turritellas* (Turritella carinata? Lea, Heilpr.), the other species found in it being comparatively few in number.

The bed has been traced by Dr. E. A. Smith from near Turnbull, in Monroe County, westward almost to the Mississippi line. Numerous localities have been examined, the list below gives the species from each. In some cases, the exact bed is not designated in the collections from Wood's Bluff; these are given in a separate column. "Wood's Bluff upper" is the green sand-bed, two to four feet thick; "Wood's Bluff lower" that part below the five to eight feet of barren clays shown in section.

	Knight's Branch.	Cave Branch.	Wood's Bluff.	Wood's Bluff (upper).	Wood's Bluff (lower).	Choctaw Corner.	Bethel, Ala.	4 miles south Mt. Sterling.	Hendrick's Marl Bed.	Butler, Choctaw Co., Ala.	½ M. S. of Butler
Ancillaria subglobosa, Con	†		†	*	*	*	*			*	*
" staminea, Con				*	*	*				*	
Natica ætites, Con	†	†		*	*	*	*	*		*	*
" Mississippiensis, Con			†		*	*	*	*	*	*	*
" limula? Con.			†	*	*	*	*	*	*	*	*
" sp					*	*				*	
" magno-umbilicata, Lea					*	*				*	
" sp					*	*					
" gibbosa, Lea										*	
" sp										*	*
" n. sp						*					
Sigaretus (Sigaticus) Clarkeanus, Aldr						*		*			
" bilix, Con					*	*				*	
Turbinella (Caricella) Bandoni, Desh	†				*				*		
Lævibuccinum linestum, Heilpr	†		*		†						
Rostellaria (Calyptrophorus) trinodifera, Con	†			*	*	*		*	*	*	*
Solarium cupola, Heilpr	†	†	†		*	*	*	*	*	*	
" delphinuloides, Heilpr			†		*	*					
" sp						*					
" elegans, Lea., var			*								
Fusus interstriatus, Heilpr	†	†	†		*	*		*			

	Knight's Branch.	Cave Branch.	Wood's Bluff.	Wood's Bluff (upper).	Wood's Bluff (lower).	Choctaw Corner.	Bethel, Ala.	4 miles south Mt. Sterling.	Hendrick's Marl Bed.	Butler, Choctaw Co., Ala.	½ M. S. of Butler.
Fusus subtenuis, Heilpr	†	†	•	○	○					○	
" (Strepsidura) subscalarinus, Heilpr..	†	†	•	○	○			○	○	○	○
" pagodiformis, Heilpr		†	†	○						○	○
" trabeatus, Con			†	○	○	○				○	○
" eugonatus, Heilpr			†	•							
" Meyeri, Aldr			○		○						
" Tombigbeensis, Aldr					○						
" sp											
" (Strepsidura), sp.?										○	
" n. sp					○						
" n. sp					○						
" n. sp					○						
" sp					○						
Pyrula multangulata, Heilpr. (juvenis, Whitf.)		†	†	○	○					○	●
" tricostata, Desh		†		○							
Pyropsis perula, Aldr				•							
? Pleurotoma acuminata, Sowb		†		●	○	○					●
Pleurotoma moniliata, Heilpr		†	○	●	•						
" (Cochlispira) cristata, Con			†	○				○		○	
" n. sp			†								
" exilloides, Aldr					○						
" Tombigbeensis, Aldr					○						
" Tuomeyi, Aldr				○	○						
" 6 species				●	○						
" sp				●							
" 2 species				○							
" n. sp				○							
Dentalium microstria, Heilpr		†	†	○	○	○	○	●	○	○	○
Cadulus, sp				○	○	○	○	○		●	
Turritella carinata? Lea		†	†	○	○	○	○	○	○	○	○
Cassidaria dubia, Aldr		†		○	●						
Voluta (Athleta) Tuomeyi, Con		†	†	○	○	○	●	○	○	●	
" petrosa? Con					○				○		○
Cancellaria evulsa, Brander (tortiplica?Con.)			†	•	○	○					
" sp					○	○					
Pseudoliva scalina, Heilpr			†							○	
" vetusta, Con			†	•	●					○	
Ranella (Argobuccinum) Tuomeyi, Aldr					●						
Trochita trochiformis, Lam					●	○		○	○	○	
Phorus reclusus? Con					○	●					
Melanopsis Choctawensis, Aldr										○	●
Turbonilla (Chemnitzia) trigemmata, Con				○	○						
Trophon gracilis, Aldr					○						
Cerithium Tombigbeense, Aldr					●						
Bulla Aldrichi, Langdon					●	○					
" n. sp						○					

	Knight's Branch.	Cave Branch.	Wood's Bluff.	Wood's Bluff (upper).	Wood's Bluff (lower).	Choctaw Corner.	Bethel, Ala.	4 miles south Mt. Sterling.	Hendrick's Marl Bed.	Butler, Choctaw Co., Ala.	½ M. S. of Butler.
Cylichna galba, Con...........................				*				*		*	
Actæon punctatus, Lea................				*							
Tornatella (Tornatellæa) bella, Con.	†		*	*	*		*	*	*	*	
Tuba antiquata, Con...				*						*	
Delphinula, sp....				*	*						
Fissurella Claibornensis? Lea.............			†								
Capulus complectus, Aldr........			†	*							
Pisania? dubia, Aldr........................			*								
Odostomia, sp.........				*	*						
Cornulina armigera, Con....				*						*	
Oliva bombylis, Con., var.....................				*				*		*	
Nassa cancellata, Lea.......................				*	*					*	
Eulima, 2 sp.				*	*					*	
Orbis rotella, Lea............................							*				
Ringicula biplicata? Lea.......................										*	
Hipponyx, sp...										*	
Clavella, sp......				*							
Columbella, n. sp..........................				*							
Scalaria carinata? Lea...				*							
Crepidula lirata, Con........				*							
Astarte tellinoides, Con........................	†			*	*			*			
" Nicklinsii, Lea., var...................				*				*			
Cytherea perovata? Con........	†			*	*	*		*	*	*	*
" Nuttalliopsis, Heilpr..............	†		*	*	*	*					
" minima, Lea.........	†			*							
? Cardita alticosta, Con........	†		*	'	*		*	*		*	*
Venericardia planicosta, Lam.................			*	*	*	*				*	*
Hippagus isocardioides, Lea...............				*							
Corbula Aldrichi, Mr......	†		*	*							?
" n. sp........				*							
Ostrea (*probably O. Thirsæ Gabb.*).......	†	†		*			*		*		
" sp..			†							*	
Neæra prima, Aldr...........................				*				*			
Leda protexta, Con...........................		†	†	*	*				*	*	
" n. sp.........................				*	*			*		*	
" n. sp				*	*	*				*	
Pecten Poulsoni, Mor........................			†				*				
" sp........				*	*						
" (Pleuronectia) n. sp...............				*							
Modiola, sp...				*							
Nucula ovula, Lea., var.				*	*	*					
Protocardia Nicolleti, Con., var..............	†			*	*			*	*		
Lucina, sp......				*	*					*	
" subvexa, Con........................				*				*		*	
" sp.				*							
Tellinia, sp............				*	*						

	Knight's Branch.	Cave Branch.	Wood's Bluff.	Wood's Bluff (upper).	Wood's Bluff (lower).	Choctaw Corner.	Bethel, Ala.	4 miles south Mt. Sterling.	Hendrick's Marl Bed.	Butler, Choctaw Co., Ala.	½ M. S. of Butler.
Avicula limula, Con....................................					*						
Pinna, sp..............................					*						
Mactra, sp..............................					*						
Diplodonta, sp........................					*						
Pteropods, 2 sp......................					*						
Cyclosmilia, sp..........	†				*	*	*	*		*	*
Madrepora, sp............................					*						
Otolithus, sp............................					*						
Nodosaria, sp					*						
Foraminifera, 2 sp.............					*						

A small layer twenty-five feet above the Wood's Bluff beds was found to contain the following species:

Cytherea perovata? Con.
Corbula Aldrichi, Mr.
Lævibuccinum lineatum, Heilpr.

Rostellaria trinodifera, Con.
Leda, n. sp. (same as W. B.).
Nassa cancellata, Lea.

TABLE VI.

BELL'S LANDING GROUP.

This group includes a large marl-bed at Bell's Landing, Ala., also a smaller bed called "Gregg's Landing Marl," which is found at the base of the bluff at Bell's Landing and in the Gregg's Landing localities. Several bluffs occur where both are exposed in one profile—on the Alabama River, notably "Lower Peach-Tree Landing;" the Bell's Landing bed occurs on the Tombigbee River, and at Tuscahoma Landing, Ala. The tables give the following localities: "Bell's Landing bed," or "Bell's Upper;" "Bell's Lower," equal to "Gregg's Landing Marl;" "Gregg's Landing Marl," equal to "Gregg's;" "Lower Peach-Tree," equal to "Bell's" and "Gregg's," together, and "Tuscahoma," equal to "Bell's Landing bed."

The locality at Bell's Landing seems to have been favorable to the growth of large specimens. To give an idea of this, the dimensions attained by the following species have been taken ; measurements are given in inches and tenths:

NAME.	Length.	Breadth.
Venercardia planicosta, Lam	5.0	4.7
Ostrea compressirostra, Say., upper valve	5.5	6.5
" " " lower valve	6.0	6.5
Turbinella pyruloides, Con.	3.9	2.5
Voluta Newcombiana, Whitf	4.7	2.0
Rostellaria trinodifera, Con	2.8	1.3
Pseudoliva scalina, Heilpr	4.3	2.7
Athleta Tuomeyi, Con	2.4	1.4
Turritella præcincta, Con	3.3	1.1
" bellifera, Aldr	3.7	0.7
? Fusus trabeatus, Con	2.3	1.2
Pseudoliva vetusta, Con.	2.1	1.6
Bulbifusus Tuomeyi, Aldr	2.8	1.9
Crassatella tumidula, Whitf..	2.5	2.2
Pectunculus stamineus, Con	2.0	2.1

	Belle Upper.	Belle Lower.	Gregg's Landing, Marl.	Lower Peach Tree.	Tuscahoma Landing.
Voluta Newcombiana, Whitf	*				
" (Athleta) Tuomeyi, Con	*				
" Sayana, Con., var	*	*	*	*	*
Fusus trabeatus, Con	*				*
" pagodiformis, Heilpr	*	*	*		*
" " var	*				
" subtenuis, Heilpr	*	*	*	*	
" subscalarinus, Heilpr	*		*	*	
" sp	*		*		
" sp	*				
" (Strepsidura), sp	*	*			
" " "					*
" spiniger, Con					*
" Meyeri, Aldr				*	
" n. sp			*		
" rugatus, Aldr			*		
Pleurotoma terebralis, Lam	*	*	*	*	*
" sp	*	*	*	*	*
" nasuta, Whitf	*	*	*	*	*
" sp	*		*		
" n. sp		*	*		*
" n. sp. (smooth)			*		*
" capax, Whitf		*	*		*
" sp		*			
Turritella Mortoni, Con	*	*	*		*
" præcincta, Con	*		*		

	Bell's Upper.	Bell's Lower.	Gregg's Landing, Marl.	Lower Peach Tree.	Tuscahoma Landing.
Turritella bellifera, Aldr..........	*		*	*	*
" multilira, Whitf..............	*		*		
" eurynome, Whitf..........	*				
Mesalia vetusta, Con., var............				*	
Potamides Alabamiensis, Whitf...........	*	*	*	*	*
Pseudoliva scalina, Heilpr.............	*				
" vetusta, Con............	*				*
" elliptica, Whitf...........	*				
" tuberculifera, Con.........	*				
Turbinella baculus, Aldr.............	*				*
" pyruloides, Con........	*				
Bulbifusus plenus, Aldr..........	*				
" Tuomeyi, Aldr............	*		*		*
Pyrula juvenis, Whitf.............	*	*		*	*
" tricostata, Desh. (Heilprin)........	*				
Fulgur triserialis, Whitf..............	*		*		
Cassidaria dubia, Aldr..........	*		*		
Fasciolaria pergracilis, Aldr.............		*	*	*	*
Pyropsis perula, Aldr...............			*		*
Melongena, n. sp.........	*				
Pisania, n. sp............	*				
" sp.........		*	*		
Cypræa Smithii, Aldr...........			*		
Rostellaria trinodifera, Con............	*	*	*	*	*
Aphorais gracilis, Aldr........	*		*	*	
Trophon, sp..........	*				
Murex, 2 sp...........	*		*		
" engonatus? Con...........			*		
Triton, n. sp...........			*		
" autopsis, Con............			*		
" exilis, Con.............			*		
" n. sp............			*		
Trochita trochiformis, Lam...........	*		*		*
Nassa cancellata, Lea..........		*	*		*
Natica erecta, Whitf...........	*				
" Alabamensis, Whitf...........	*	*	*	*	*
" aperta, Whitf.........	*	*	*	*	*
" onusta, Whitf...........	*	*	*		*
" perspecta, Whitf.............	*				
" parva, Lea., var.........	*				
" decipiens, Mr.............	*				
" Mississippiensis, Con., umbilicated var...........	*		*		
" sp.............	*	*	*		*
Sigaretus bilix, Con...............	*				
" declivis, Con............	*				
Cancellaria, sp............	*		*		
Oliva gracilis, Lea...........	*		*		
Cadulus, sp............	*				
" 3 sp...........			*		*
Dentalium microstria, Heilpr............	*				
Scalaria, n. sp.............			*		

	Bell's Upper.	Bell's Lower.	Gregg's Landing Marl.	Lower Peach Tree.	Tuscahoma Landing.
Adeorbis depressus, Lea	*				
Teinostoma subrotunda, Mr	*				
Solarium elegans, Lea, var				*	
" delphinuloides, Heilpr			*		
" scrobiculatum, Con			*		
Velutina expansa, Whitf	*		*		
Cylichna galba, Con	?		*		
Bulla, n. sp			*		
" (Haminea) Aldrichi, Langdon	*		*		
Scaphander, n. sp			*		
Tornatina? sp	*				
Odostomia, sp	*				
Chemnitzia trigemmata, Con			*		
" n. sp			*		
Eulima notata, Lea			*		
" ? sp			*		
Melanopsis, n. sp..*			*		
" anita, Aldr			*		
Ringicula, sp	*				
Cytherea Nuttalliopsis, Heilpr	*	*	*		*
" perovata, Con	*				
Ostrea compressirostra, Say	*				
" sp.?	*		*		
Pecten, n. sp	*				
" Deshayesii, Lea			*	*	
Cardium Nicolleti, var			*		
Modiola, sp			*		
Avicula limula, Con			*		
Crassatella tumidula, Whitf	*		*		
" sp			*	*	
Dosinopsis lenticularis, Rogers	*				
Lucina compressa, Lea	*				
" pomilia, Con	*				
" Claibornensis, Con			*		
" sp			*		
Diplodonta, sp			*		
Tellina, 4 sp			*		
Arca lima, Con., var	*		*		
Cucullæa transversa, Rogers		*	*		
Venericordia planicosta, Lam	*		*		*
" rotunda, Lea., var	*		*	*	*
Pectunculus stamineus, Con	*	*	*	*	
" " var	*				
Pholas alatoidea, Aldr	*		*		*
Egeria inflata, Lea	*				
" plana? Lea	*		*		*
" subtrigonia, Lea	*				
" rotunda, Lea				*	
Leda protexta, Con	*		*	*	
" sp			*	*	*

	Bell's Upper.	Bell's Lower.	Gregg's Land ing. Marl.	Lower Peach Tree.	Tusaboma Landing.
Martesia elongata, Aldr.	*				
Panopæa, sp.			*		*
Lithodomus Claibornensis, Con.	*				
Corbula engonata, Con.	*		*	*	*
" n. sp.	*				
" Aldrichi, Mr.			*		*
Nucula magnifica, Con.	*		*		
Cyclosmilia, sp	*	*	*	*	*
Massive Coral.			*	*	
Nodosaria, sp.	*			*	

TABLE VII.

NANAFALIA GROUP.

The most remarkable fossil of this division is *Gryphæa thirsæ*, Gabb., which is abundant through sixty feet of strata. The best location for its fossils is near the base under the *Gryphæa* bed. As nearly all the specimens on hand are from the exposures on the Tombigbee River, no other localities are mentioned.

Turritella Mortoni, Con.
"   bellifera, Aldr.
"   n. sp.
Pseudoliva vetusta, Con.
"   scalina, Heilpr.
"   erecta, Aldr.
Ancillaria subglobosa, Con.
"   n. sp.
Rostellaria trinodifera, Con.
Voluta Sayana, Con.
"   Newcombiana, Whitf.
"   (Athleta) Tuomeyi. Con.
Volutalithes, sp.
Fusus trabeatus, Con.
"   n. sp.
"   sp.
Exilia pergracilis, Con.
Pleurotoma, sp.
"   persa, Whitf.
"   sp.
"   terebralis, Lam.
Natica eminula, Con.
"   ætites, Con.
Dentalium thalloides, Con.

Dentalium sp. (smooth).
Cadulus, sp. ———
Melanopsis Choctawensis, Aldr.
"   n. sp.
Ringicula, n. sp.
Odostomia, sp.
Bulla, sp.

Venericardia planicosta, Lam.
Gryphæa thirsæ, Gabb.
Pectunculus stamineus, Con.
Crassatella tumidula, Whitf.
Panopæa, sp.
Arca, n. sp.
Cardium Tuomeyi, Aldr.
Corbula Aldrichi, Mr.
Tellina, sp.
Nucula ovula, Lea.
Leda, sp.
Cytherea Nuttalliopsis, Heilpr.
Pinna, sp.
Ostrea compressirostra, Say.

Cyclosmilia? sp.

TABLE VIII.

MATTHEWS' LANDING GROUP.

The exposure from which this collection was made is on the Alabama River, about seven miles below Prairie Bluff. One other point in Wilcox County, on Mr. Jones' place (S. 12, T. 12, R. 6 E.), has the identical fossils. The list is made from the typical deposit. Nearly all the univalves are remarkable for their ornamentation. *Rostellaria velata*, Con., absent from all the other groups below the "Claibornian," appears once more here.

Murex morulus, Con.
"   Matthewsensis, Aldr.
Triton Showalteri, Con.
?   "   n. sp.
Strepsidura, sp.
Fusus tortilis, Whitf.
"   Meyeri, Aldr.
"   6 species.
"   pagodiformis, Heilpr.
Exilia pergracilis, Con.
Neptunea Matthewsensis, Aldr.
Leucozonia biplicata, Aldr.
Pleurotoma persa, Whitf.
"   adeona, Whitf.
"   12 species.
Trochus Alabamensis, Aldr.
Solarium, sp.
"   n. sp.
Pseudoliva unicarinata, Aldr.
"   vetusta, Con.
"   scalina, Heilpr.
Ancillaria staminea, Con.
Rostellaria velata, Con.
Pyrula juvenis, Whitf.
Volutalithes limopsis, Con.
"   rugata, Con.
Voluta Showalteri, Aldr.
Lævibuccinum lineatum, Heilpr.
Natica perspecta; Whitf.
"   reversa, Whitf.
"   Alabamensis, Whitf.

Natica eminula, Con.
"   n. sp.
Sigaretus? n. sp.
Melanopsis Choctawensis, Aldr.
Cadulus turgidus, Mr.
Dentalium microstria, Heilpr.
Cerithiopsis, n. sp.
Odostomia, sp.
Eulima, sp.
Rissoina, n. sp.
Ringicula, n. sp.
Cylichna, sp.
Turritella humerosa? Con.
"   Alabamensis, Whitf.
"   multilira, Whitf.
"   Mortoni, Con.

Cucullæa macrodonta, Whitf.
Pecten (Pleuronectia) Alabamensis, Aldr.
Venericardia rotunda, Lea, var.
"   Sillimani, Lea.
Corbula, sp.
Leda eborea, Con.
Nucula magnifica, Con.
Cardium, sp.
Avicula limula, Con.
Astarte, sp.

Nodosaria, sp.
Coral, sp.

TABLE IX.

BLACK BLUFF GROUP.

This bed consists principally of a bed of very dark clay, which has a fine exposure on the Tombigbee River at Black Bluff, Sumpter County, and I believe is the lowest Tertiary seen on that river. It forms a belt of low-lying prairie just south of the Cretaceous hills.

Its fossils would unite it to the Matthews' Landing group.

Nautilus, n. sp. (fragments).
Exilia pergracilis, Con.
Mitra, n. sp.
Aphorais gracilis, Aldr.
Ancillaria, sp.
Pleurotoma, 2 sp.
Fusus, 2 sp.
Cadulus turgidus, Mr.
Volutalithes rugata, Con.
Dentalium, sp.

Tornatella bella, Con.

Pecten (Pleuronectia) Alabamensis, Aldr.
Cucullæa macrodonta, Whitf.
Nucula maghifica, Con.
Leda, n. sp.

Coral, sp.
Crab remains (abundant).
Foraminifera, 2 sp.

TABLE X.

MIDWAY GROUP.

Consists of a few feet of material at the base of the Tertiary ; a bed containing a species of Turritella being immediately contiguous to the Cretaceous. The fossils of this and the previous group are either distorted or badly preserved, and by weathering of the materials often mixed with the underlying Cretaceous. The species here mentioned were all found "*in situ.*"

1st. MIDWAY BED consisting of a hard, silicious limestone.

Enclimatoceras Hyatti? White.

2d. TURRITELLA ROCK.

Turritella Mortoni? Con.
"   humerosa? Con.
Venericardia planicosta, Lam.

Ostrea, 2 sp.
Coral, 2 sp.

In conclusion, it must be understood that these lists represent collections made personally in company with Dr. E. A. Smith and D. W. Langdon, Jr., and are as nearly accurate as the writer can make them, without a series of fossils carefully compared with types ; they represent many days and nights of hard work.

FINIS.

Explanation of Plate I.

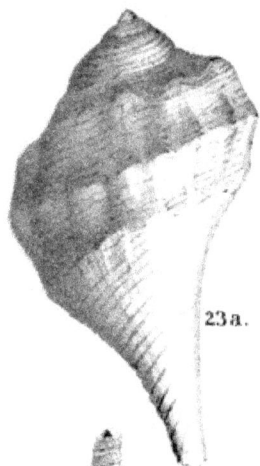

23 a.

15

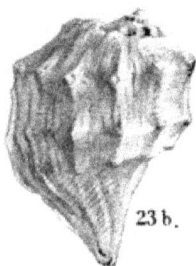

23 b.

14

13.

18.

22

16.

19 b.

17 a.   17 b.

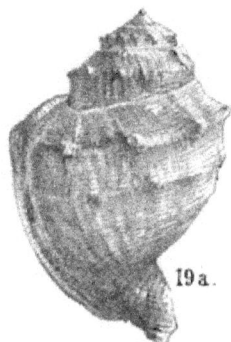

19 a.

20.

21.

Explanation of Plate II.

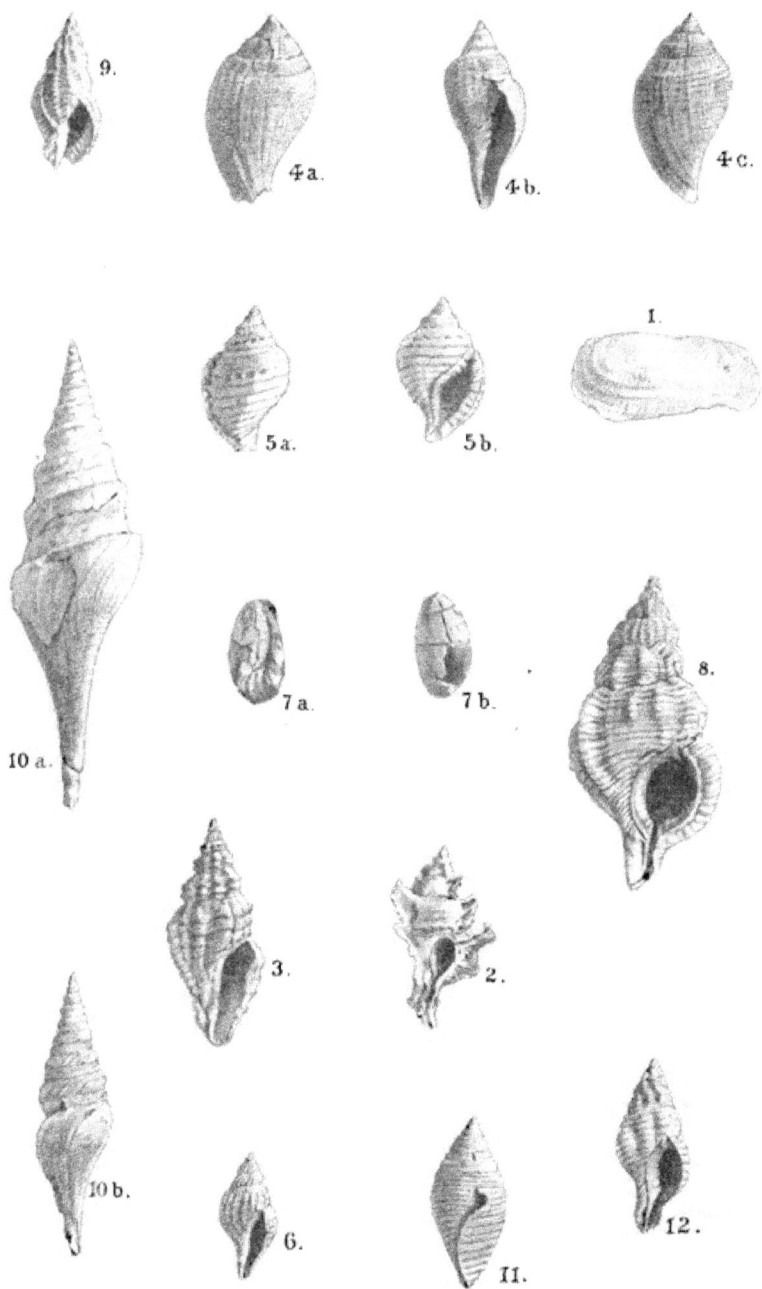

Explanation of Plate III.

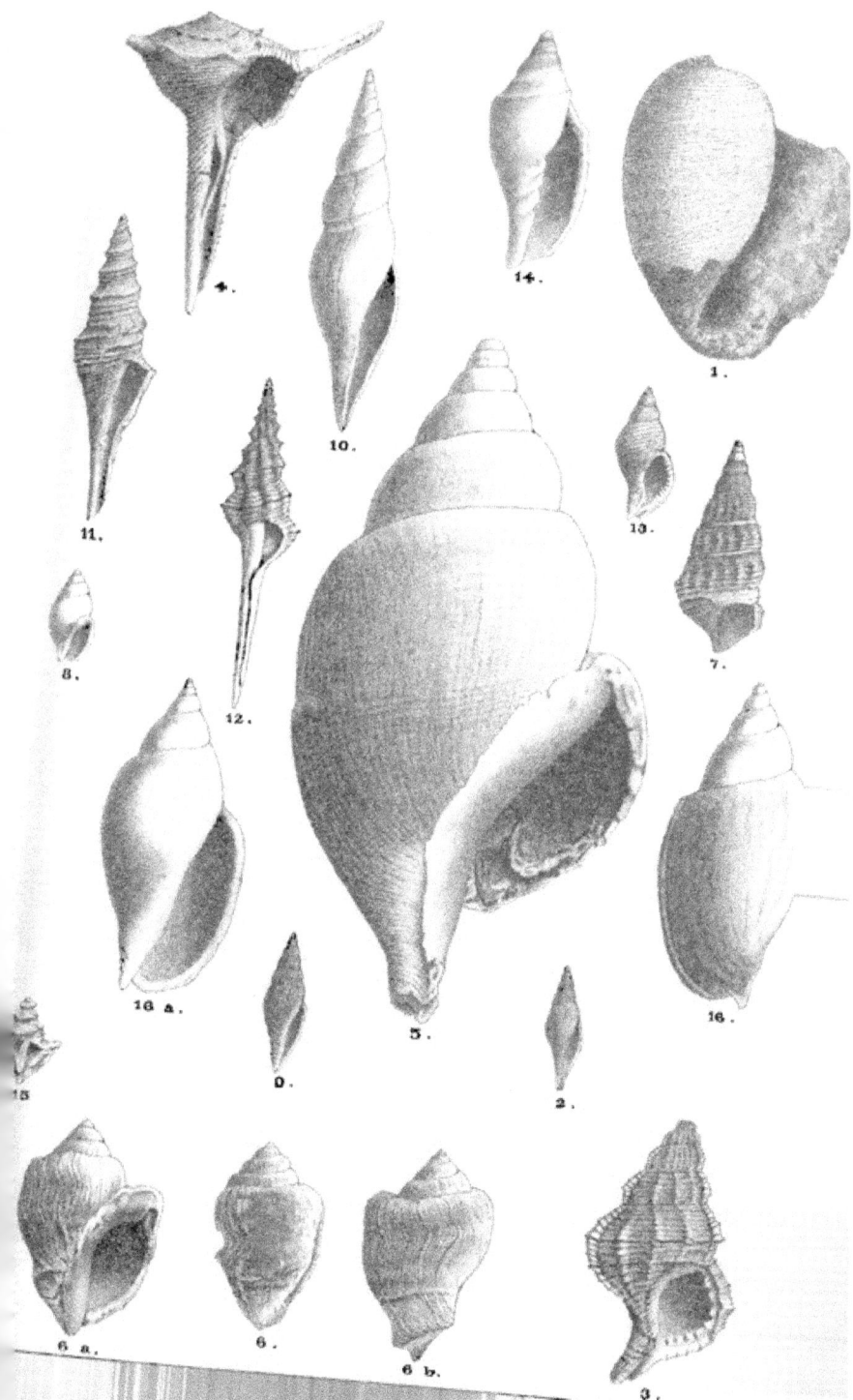

1. 2. 3. 4. 5. 6 a. 6. 6 b. 7. 8. 9. 10. 11. 12. 13. 14. 15. 16. 16 a.

Explanation of Plate IV.

By mistake of the lithographers this figure is numbered 4 instead of 9.

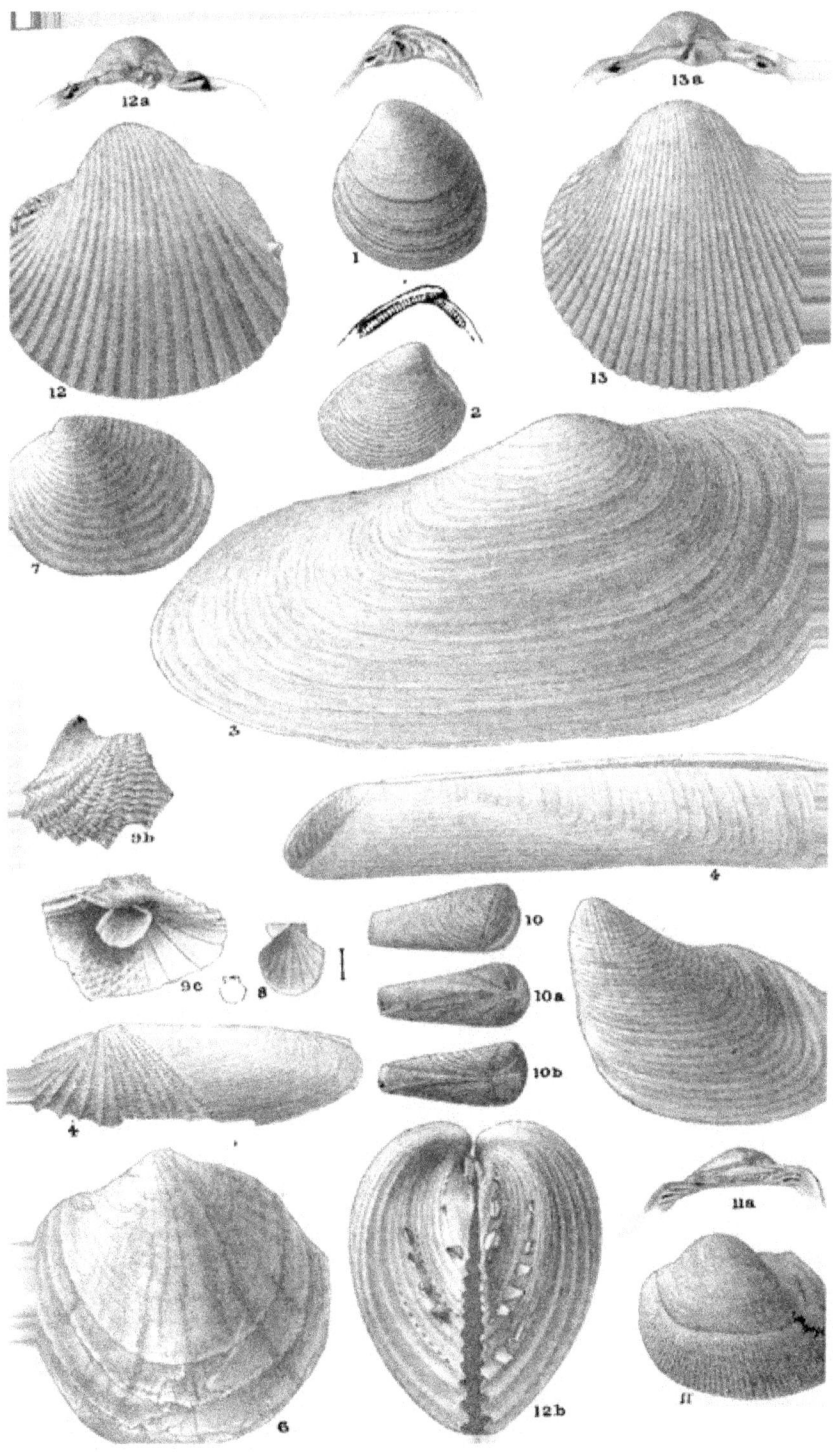

Explanation of Plate V.

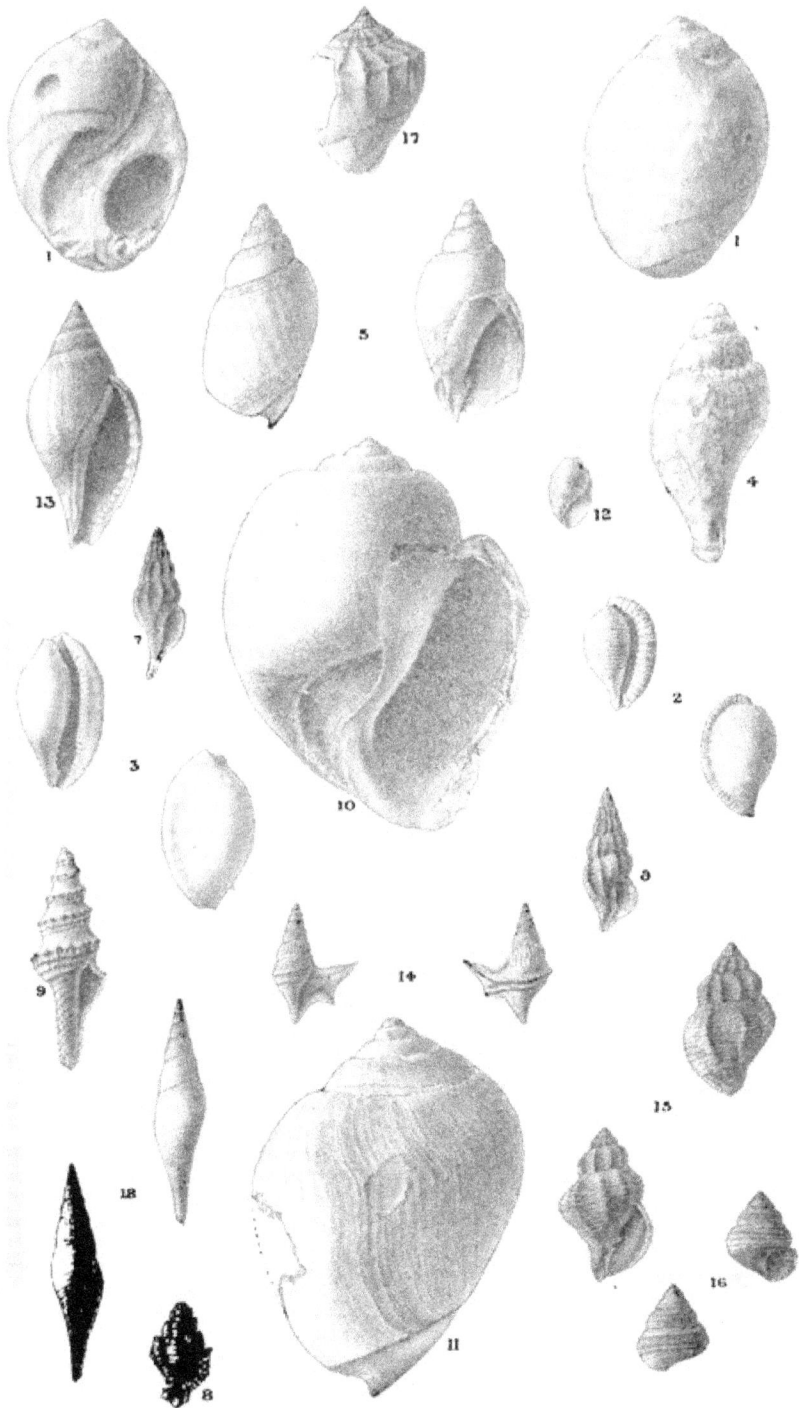

Explanation of Plate VI.

II.

CONTRIBUTIONS

TO THE

Eocene Paleontology of Alabama and Mississippi.

BY

OTTO MEYER, Ph. D.

Contributions to the Eocene Paleontology of Alabama and Mississippi.

BY OTTO MEYER, Ph. D.

In the following a number of new or unfigured species of invertebrates of the Eocene Tertiary of Alabama and Mississippi are described and figured. A very few known species are refigured for some special reason. Those descriptions and figures which belong to Plate 1 were made in the year 1884, but the publication of this plate has been delayed by various reasons. I am under obligation to Mr. T. H. Aldrich and Dr. E. A. Smith for the present opportunity to publish this paper. The type specimens are in my collection. In the following descriptions of univalves, the term "longitudinal" means parallel to the suture or helicoid axis of the shell, "transverse" indicates the direction rectangular to it.

DENTALIUM LEAI MEYER. Pl. 1, figs. 2, 2a. *Am. Jour. Sci.*, 1885, XXIX., p. 462.

?*Dentalium arciformis*, Con. *Am. Jour. Sci.*, 1, 2d ser., 1846, p. 212. Pl. 1, fig. 3.

Substance of the shell thick, slightly curved, smooth ; section and aperture circular ; smaller termination notched in the direction of the curvature ; interiorly with a tube.

Locality.—Claiborne, Ala.

Three specimens have the incision of the smaller termination, but only one of them is perfect enough to show the interior tube also. A young specimen approaches in its aperture the following species :

The smooth species of Dentalium have to be distinguished by their aperture. The aperture is wanting in Conrad's specimen of *Dentalium arciformis*. Later, Conrad withdrew this species; at least he did not mention it in his lists, probably considering his form a synonym of *Dentalium turritum*, Lea,* which species, however, in my opinion, is a fragment of a Cadulus.

DENTALIUM DANAI, Meyer.   Pl. 3, figs. 2, 2a.   *Am. Jour. Sci.*, 1885, XXIX., p. 462.

Smooth, section circular; smaller aperture with additional tube; margin distinctly notched on the convex side of the shell; slightly emarginate on the concave side

Locality.—Jackson, Miss.

The preceding species is notched distinctly on both sides of the margin of the aperture.

DENTALIUM SUBCOMPRESSUM, Meyer.   Pl. 3, figs. 3, 3a.   *Am. Jour. Sci.*, 1885, XXIX., p. 462.

Shell small, smooth, somewhat polished; section ovate.

Locality.—Jackson, Red Bluff, and Vicksburg. Miss.

The figured type-specimen is from Jackson.

DENTALIUM BITUBATUM, n. sp.   pl. 3, fig. 1.

Smooth, rapidly increasing in size; section suborbicular; aperture with a long additional tube.

Locality.—Jackson, Miss.

I have only one specimen of this species, which in its long tube of the aperture resembles *Dentalium duplex*, Def.† from the Paris basin.

DENTALIUM ANNULATUM, n. sp.   Pl. 1, fig. 1.

Small, slightly tapering; section circular; surface annulated.

Locality.—Claiborne, Ala.

I found only the figured specimen. Without a magnifying glass the characteristic rings of the surface are scarcely to be seen. It is very similar to young specimens of *Dentalium minutistriatum*, Gabb.,‡ which species is annulated on its youngest part, a fact not mentioned by Gabb.

---

* I. Lea, *Contrib. to Geol.*, 1833, p. 35, pl. 1, fig. 3.

† Desh., An. s. vertèb., II., p. 203, pl. 1, figs. 36–39.

‡ *Jour. Acad. Nat. Sci.*, Phila., IV., 2d ser., 1860, p. 386, plate 67, fig. 46.

In the Claiborne specimen, however, the rings are larger and of a different appearance, and there is no trace of longitudinal striæ.

CADULUS VICKSBURGENSIS, Meyer. Pl. 3, fig. 6. *Am. Jour. Sci.*, XXIX., 1885, p. 463.

Inflation very faint near the end, somewhat compressed ; smaller aperture with four turret-like appendages, one opposite pair of which is broader than the other pair.

Locality.—Vicksburg, Red Bluff, Miss.

The type specimen is of the "Higher Vicksburgian." The species is distinctly compressed, but less than Cadulus depressus, Mr., from Claiborne.

CADULUS QUADRITURRITUS, n. sp. Pl. 3, figs. 7, 7a.

Not compressed ; inflation near the end distinct but gradual ; smaller aperture with four equal rounded appendages, divided by notches of the same shape.

Locality.—Red Bluff, Miss.

CADULUS JACKSONENSIS, Meyer. Pl. 3, figs. 8, 8a, 8b. *Am. Jour. Sci.*, XXIX., 1885, p. 462.

Rather large ; inflation faint near the end, very slightly compressed ; smaller aperture elliptical ; margin by notches divided into four appendages ; the two appendages on the smaller side of the ellipse are slender, simple and equal to each other ; the two other opposite ones are broad, emarginate in the middle, and unequal in size, that one situated on the convex side of the shell being the largest.

Locality.—Jackson, Miss.

CADULUS TURGIDUS, n. sp. Pl. 1, fig. 10.

Width of the shell rapidly increasing for about two-thirds of the entire length, and then more rapidly decreasing ; section circular.

Locality.—Matthews' Landing, Ala.

Rather common ; I received this shell from Mr. Aldrich. It differs by its very strong inflation from all the other species of Cadulus of the Southern Tertiary which I know of.

CADULUS CORPULENTUS, n. sp.   Pl. 3, fig. 5.

Small; inflation near the middle, short and stout, not compressed; smaller aperture elliptical with simple margin.

Locality.—Red Bluff, Miss.; common.

CADULUS JUVENIS, n. sp.   Pl. 3, fig. 4.

Small; inflation near the middle; slender, not compressed; smaller aperture forming an ellipse, one side of which is flattened.

Locality.—Jackson, Miss.; not rare.

TEINOSTOMA ANGULARIS, Meyer.   Pl. 1, figs. 9, 9a.   *Am. Jour. Sci.*, 1885, XXIX., p. 463.

Lenticular; whorls three, rapidly increasing in size, last whorl carinated; base slightly convex, nearly flat; columellar part thickened; mouth rhombical.   On both sides of the carina several impressed spiral lines crossed by very minute radiating striæ.   Last whorl with an indistinct spiral fold near the suture.

Locality.—Claiborne, Ala.

TEINOSTOMA SUBROTUNDA, Meyer.   Pl. 2, figs. 26, 26a.   *Am. Jour. Sci.*, 1885, XXIX., p. 463.

Discoid; margin rounded; umbilical region covered and thickened by callus; whorls four, rapidly increasing in size; smooth, except some faint revolving lines; suture indistinct; aperture quadrate-elliptical.

Locality.—Claiborne, Ala.

TEINOSTOMA VERRILLI, Meyer.   Pl. 2, figs. 27, 27a.   *Am. Jour. Sci.*, 1885, XXIX., p. 463.

Discoid; umbilical region covered and thickened by callus; margin angular, though not carinated, polished; suture entirely indistinct, so that the number of whorls can not be counted; base regularly rounded; aperture trigonal-elliptical.

Locality.—Jackson, Miss.

Related to the preceding species, but distinguished by the more angular margin, and the entire absence of sculpture.   The first difference implies also the different form of base and aperture; the second difference, the indistinctness of the suture.

ADEORBIS SUBANGULATUS, n. sp.   Pl. 2, fig. 28.

Discoid; whorls five, rapidly increasing in size; margin somewhat angular; basal part of margin rounded; umbilicus deep; suture distinct; surface with revolving lines, indistinct near the margin; aperture irregularly elliptical.

Locality.—Jackson, Miss.

Adeorbis depressus, Lea., sp. (Teinostoma rotula, Heilpr.) from Claiborne has the umbilicus nearly closed, a regularly rounded margin, a more developed ornamentation, and is larger.

ADEORBIS LAEVIS, n. sp.   Pl. 2, figs. 29, 29a.

Discoid; umbilicus large; margin rounded; aperture circular; whorls five; convex, smooth; suture distinct.

Locality.—Red Bluff, Miss.

In Vicksburg there occurs a variety "var Vicksburgensis," with smaller umbilicus.

SOLARIUM HARGERI, n. sp.   Pl. 2, figs. 23, 23a, 23b.

Discoid; surface flattened; base convex; margin carinated; whorls five. The ornamentation of the older whorls consists of four granulated revolving striae, the two in the middle smaller than those near the suture. Suture canaliculated; base with revolving lines, those near the umbilicus crenulate; aperture rhombical.

Locality.—Red Bluff, Miss.

The canaliculation along the suture is the main characteristic of this species. Named in honor of Mr. Oscar Harger, the able and careful naturalist in New Haven, Conn.

SCALARIA GRACILIOR, n. sp.   Pl. 2, fig. 2.

Small, subulate; nucleus consisting of two rounded, spirally striated whorls. Five adult whorls are regularly rounded and covered with regular, sharp, transverse ribs, which extend also over the base. Aperture elliptical.

Locality.—Claiborne, Ala.

I found only the figured specimen.

EGLISIA PULCHRA, n. sp.   Pl. 1, fig. 16.

Whorls convex, regularly rounded, covered with five flat longitudinal lines, diminishing in breadth from the top to the base of each whorl. They are crossed by numerous fine elevated striæ, which are broad and indistinct, or obsolete on the spiral lines, but sharp and distinct in the interstices.

Locality.—Claiborne, Ala.

In its sculpture the species is remarkably similar to *Scalaria (Eglisia) vincta*, Desh.,* but it is much larger.

EGLISIA REGULARIS, n. sp.　Pl. 2, fig. 3.

Whorls rather rapidly increasing in size, regularly rounded; covered by four sharp elevated revolving lines, the uppermost of which is the smallest; the interstices with numerous transverse riblets, which in connection with the revolving lines give to the surface an almost cancellate appearance.

Locality.—Red Bluff, Miss.

EGLISIA INAEQUISTRIATA, n. sp.　Pl. 2, fig. 4.

The only specimen found, a fragment of two whorls, has less rounded whorls than the preceding species; they are covered by six revolving striæ, alternating in size and far less elevated than in *Eglisia regularis;* the little riblets in the interstices are similar to those in the preceding species; aperture rounded; subeffuse anteriorly; base flattened, spirally striated.

Locality.—Red Bluff, Miss.

CÆCUM SOLITARIUM, n. sp.　Pl. 3, fig. 9.

Cæcum, sp., *Am. Jour. Sci.,* XXX., 1885, p. 71.

Small, regularly curved, somewhat contracted at the aperture; section and aperture circular; smooth, except concentric rings of growth.

Locality.—Vicksburg, Miss.; "Lower Vicksburgian."

CRUCIBULUM ANTIQUUM, n. sp.　Pl. 1, fig. 11.

Subconical; margin oval, striate within; diaphragm entire; rhombical, close to the shell.

Locality.—Claiborne, Ala.

The surface of the single specimen is badly preserved　If I am not mistaken it is the first Crucibulum found in the Old Tertiary Formation.

---

*Desh., An. s. vertèb., II., p. 353, pl. 23, figs. 17, 18, 19.

NATICA DECIPIENS, Meyer. Pl. 2, fig. 22. *Am. Jour. Sci.*, XXIX., 1885, p. 464.

Spire rather elevated; whorls six, flattened; aperture semilunar, two-thirds of the entire length; umbilicus deep; on the inner lip callus is spread, which forms a prominence at the posterior end of the aperture; inner lip distinctly emarginate in front of the umbilicus.

Locality.—Vicksburg, Miss.

It has a more transverse form than *Natica parva*, Lea., from Claiborne, which is else very similar. The type-specimen is from the "Lower Vickburgian"; but it occurs also in the "Higher Vicksburgian."

AMAURA TORNATELLOIDES, n. sp. Pl. 1, fig. 12.

Shell oval; imperforate, smooth; spire conical; apex obtuse; suture very distinct; whorls five; columella thickened; aperture semicircular, acute posteriorly, and with an indistinct broad emargination anteriorly.

Locality.—Claiborne, Ala.

I found only the figured specimen.

RISSOINA MISSISSIPPIENSIS, n. sp. Pl. 2, fig. 17.

The only specimen found has four smooth, embryonic whorls; the first two of which form a disk, thus making the apex blunt; then four adult whorls follow which are slightly convex, and are densely covered by indistinct, rounded, oblique, transverse ribs; base with faint revolving lines; aperture semilunar, faintly channeled anteriorly; receding posteriorly, thus giving a strongly sigmoid appearance to the outer lip.

Locality.—Jackson, Miss.

ACLIS MODESTA, n. sp. Pl. 2, fig. 1.

Minute; the sinistral nearly vertical; nucleus consists of two smooth whorls; adult whorls five; little convex; flattened immediately below the suture; covered with microscopical spiral lines; aperture regularly elliptical.

Locality.—Claiborne, Ala.

Only the type-specimen was found.

TURBONILLA NEGLECTA, n. sp. Pl. 1, fig. 4.

Subulate; whorls flattened, closely covered with broad and distinct transverse ribs, ending rather abruptly on the last whorl and leaving the base smooth; aperture subquadrangular, longer than broad; inner lip

twisted, thus forming an obtuse fold at some distance from the sharp
outer lip; **crenulate within by a** few distant **elevated** spiral lines.
Locality.—Claiborne, Ala.

TURBONILLA MISSISSIPPIENSIS, n. sp   Pl. 2, **fig. 5.**

Subulate; the sinistral nucleus consists of three smooth convex whorls,
which very rapidly increase in size; its axis is horizontal. Seven convex adult whorls are covered with elevated, tranverse ribs, about twelve
on each whorl. Suture distinct, aperture oval, base smooth.
Locality.—Red Bluff, Miss.

Differs from the preceding species mainly by its convex whorls.

ODOSTOMIA BIDENTATA, n. sp.   Pl. 1, fig. 3.

Subulate, whorls eight; nucleus and first whorl smooth, elsewhere furnished with transverse ribs; suture impressed; base rounded, smooth;
columella with a very prominent oblique fold, and a second indistinct one
below it.
Locality.—Claiborne, Ala.

The ribs have in one of the three specimens a tendency to become
obsolete, especially toward the lower part of each whorl. The lower fold
will be better seen in fragments, with exposed columella, than in a perfect mouth.

CHEMNITZIA ACUTA, n. sp.   Pl. 2, fig. 6.

Whorls eleven, nearly flat; aperture ovate   The ornamentation consists of sharp, transverse ribs, crossed by less prominent spiral lines.
Each of the four smooth, embryonic whorls is much smaller than the
first ornamented whorl, thus forming a pointed nucleus; suture distinct.
Locality.—Red Bluff, Miss.

Chemnitzia Claibornensis, Heilpr., sp.,* has the revolving **lines not**
elevated, but impressed; and the transverse ribs are not sharp **lines, but**
obtuse folds

BITTIUM KOENENI, n. sp.   Pl. 2, fig. 12.

Four spirally striated, embryonic whorls without ribs are followed by
four transversely ribbed whorls with three spirals, the uppermost of which
is the smallest. On the following older whorls, three more spirals appear

---

Proc. Ac. Nat. Sci., Philadelphia, 1879; p. 214, pl. 13, fig. 11.

between them. All the whorls are convex ; aperture effuse anteriorly, but without proper canal ; base spirally striated. Many of the specimens have varices.

Localities.—Jackson, Miss., common ; Red **Bluff, Miss.**, not rare.

The type-specimen is from Jackson ; in the specimens from **Red Bluff** the canal is more distinct. Named after Prof. v. Kœnen, who has successfully worked up the German Tertiary.

CERITHIOPSIS ALDRICHI, n. sp. Pl. 2, fig. 14.

Subulate ; whorls convex ; oldest whorls covered by four spiral lines, the second one from above is the smallest and last developed one ; they are covered by transverse ribs, about twelve on each whorl ; the embryonic whorls are numerous and rounded—on the oldest of them the spiral lines commence to appear, while the others are only covered by numerous curved, transverse ribs. Base covered with minute elevated, revolving lines, the outermost of which is larger ; canal reflected.

Localities,—Red Bluff, Miss., Jackson, Miss., Claiborne, Ala.

CERITHIOPSIS JACKSONENSIS, n. sp.

Whorls regularly rounded, covered by four elevated, longitudinal lines, crossed by numerous transverse ribs of smaller size ; the points of crossing are thickened. Base covered with minute elevated, revolving lines, the outermost of which is larger ; canal reflected.

Locality.—Jackson, Miss.

On the last whorl of the type-specimen a fifth spiral line appears near the suture, and in a much larger specimen this fifth spiral is fully developed. Differs from the preceding species in having more rounded whorls.

TRIFORIS SIMILIS, n. sp. Pl. 1, figs. 8, 8a.

Slender ; whorls flat, covered with three longitudinal lines of nodules ; the middle line the smallest and apparently last developed, but on the lower whorls equal in size to the other two ; mouth quadrate ; base flattened, with a distinct impressed line along its margin ; suture very distinct. Length of the last five whorls, three mm.

Locality.—Claiborne, Ala.

The species is very similar to the Miocene *Cerithium moniliferum*, H. C. Lea,[*] which I do not have. Besides, it resembles very much that

---

[*] *Trans. Am. Phils. Soc.*, IX., Second Series, 1843, p. 269, pl. 37, fig. 92 ; and Emmons' *Rep. North Carolina Geol. Surv.*, p. 269, fig. 159.

species of the German Oligocene of Waldbœckelheim, which had been considered as the recent *Triforis perversus*, L., but which v. Kœnen calls *Cerithium Bœttgeri*.* I am of the opinion that this German form is a true *Triforis*. One of my larger specimens, with an outer lip more complete than is commonly the rule, has the canal nearly closed. The absence of a third opening is no absolute proof, as some recent sure species of *Triforis* (for instance, *Trif. nigro-cinctum*, Ad., of the American Coast) usually do not show it. The number of spiral lines on the base of this German species is variable, depending mainly, but not alone, upon the size of the shell. A few of the largest specimens are distinguished by a fourth line of nodules on the last whorls.

TRIFORIS MAJOR, n. sp.   Pl. 1, fig. 6.

Large ; whorls flat, lower whorls with three longitudinal lines, formed by nodules of equal number in each line ; the nodules of the uppermost spiral are the largest and touch nearly those of the middle line ; they are rounded and separate from each other ; those of the two lower spirals are compressed ; the middle is the smallest and apparently last developed ; base flattened, with a distinct impressed line along its margin ; length of the two lower whorls, five mm.

Locality.—Claiborne, Ala.

TRIFORIS MERIDIONALIS, n. sp.   Pl. 2, fig. 15.

Large ; whorls flat, with three longitudinal lines, which are continuous and only faintly nodulus—the lowest is the largest, the middle line is only one-third and the uppermost two-thirds of its breadth ; the suture is defined by one more small, elevated line ; base with a revolving line along the suture.

Locality.—Red Bluff, Miss.

TRIFORIS, sp.   Pl. 1, fig. 7.

Small ; whorls flat, with three noduliferous spiral lines, the middle of which is the smallest ; of these lines, especially the uppermost one is sometimes broken up in single nodules, but in general they are continuous ; base with radial lines of growth and a spiral along its margin.

Locality.—Claiborne, Ala.

Resembles the preceding species.

* v. Kœnen, *Norddeutsch. Miocœn.*, *Zweiter Theil*; *N. Jahrb. f. Mineralogie*, etc., 1882, Beilageband II., p. 272.

TRIFORIS DISTINCTUS, n. sp.   Pl. 1, figs. 5, 5a.

Small, slender; whorls convex, with three noduliferous, longitudinal lines, the uppermost of which is the smallest; suture marked by a small, plain, elevated spiral line; mouth subquadrate; base with a distinct and an indistinct spiral line; length of the five lower whorls, two and one-half mm.

Locality.—Claiborne, Ala.

The species is very distinct from the preceding in its convex whorls and in the line along the suture. Owing to the lowest spiral, the whorls appear almost carinated at their lowest part. An essential characteristic is in the upper line being the smallest and apparently last developed, the species resembling, in this point, the German Tertiary *Cerithium (Triforis) Fritschi*, v. Koenen.*

TRIFORIS BILINEATUS, n. sp.   Pl. 2, fig. 16.

Small; whorls flat, with two lines of separate, rounded nodules; base with a revolving line near the suture.

Locality.—Red Bluff, Miss.

Differs from all the preceding species by having only two revolving lines. As the distance of the lines is about equal, the shell looks like being uniformly covered with spirals, and the whorls can scarcely be distinguished from each other.

NASSA MISSISSIPPIENSIS, Con., sp. var.   Pl. 2, fig. 11.

Besides the typical form, there occurs both in the Higher and Lower Vicksburgian, but is rare, a variety with more numerous transverse ribs.

CANCILLARIA TURRITISSIMA, n. sp.   Pl. 1, fig. 15.

Turreted; the convex whorls very oblique to the axis of the shell; apex obtuse, smooth, the shell otherwise covered with numerous spiral lines, crossed by some transverse, large ribs; aperture semicircular; columella with three folds—the uppermost very prominent, the lowest indistinct; outer lip sharp, striate within at some distance from the aperture.

Locality.—Claiborne, Ala.

The first two embryonic whorls form almost a disk, thus making the

---

* *N. Jahrb. f. Mineralogie*, etc., 1882, Beilageband II., p. 271, pl. 6, fig. 19.

apex obtuse. A species similarly distinguished by its very slender form is *Cancellaria Bezanconi*, de Boury,* from the French Tertiary.

LATIRUS HUMILIOR, Meyer.   Pl. 2, figs. 20, 20a.

Turbinella humilior, Meyer.   *Am. Jour. Sci.*, XXIX., 1885, p. 464.

Fasciolaria Jacksonensis, Aldrich.   *Jour. Cin. Soc. Nat. Hist.*, 1885, p. 150, pl. 2, fig. 12.

Fig. 20 represents my oldest specimen; its canal is somewhat shorter than usual. Fig. 20a is a young specimen. Fig. 19 shows, for comparison, *Latirus protracta*, Con., sp., from Vicksburg. I refer to the above-cited place.

MUREX ANGULATUS, n. sp.   Pl. 2, fig. 18.

Ovate; nucleus consisting of four smooth, embryonic, rounded whorls, which rapidly increase in size; adult whorls angular, with edged, transverse ribs, which, on the angles, are produced into spines—on the body whorl they number ten; a few distinct, elevated, revolving lines cover the lower part of each whorl; canal slightly curved.

Locality.—Jackson, Miss.

MUREX SIMPLEX, Aldr., var.: *aspinosus*, n. var.   Pl. 2, fig. 21.

Ovate; whorls convex, with obtuse, transverse ribs—seven on the body whorl; covered by elevated, sharp, distinct, revolving lines, which on the older whorls alternate with faint ones; canal nearly straight; outer lip thickened, crenulate within; callus of inner lip with three obtuse teeth.

Locality.—Red Bluff, Miss.

Murex simplex, Aldr., from Byram Station, Miss., has spinous ribs.

TURRICULA CINCTA, n. sp.   Pl. 1, fig. 13.

? *Mitra gracilis*, H. C. Lea.   *Am. Jour. Sci.*, XI., 1841, p. 101, pl. 1, fig. 20.

Slender, except near the apex, where the whorls are rapidly diminishing in size; whorls eight, rather flat; except the first two, smooth, embryonic whorls, covered with transverse ribs, which become obsolete on the last whorls; parallel to the suture, an impressed spiral line produces a broad band along the suture; aperture narrow; columella with four folds, downward diminishing in size and extending over the base of the shell. striat-

---

* *Mém de la Soc. Géol.*, 3me ser., III., 1884, p. 105, pl. 3, fig. 8.

ing it; outer lip sharp, spirally striated within at some distance from the margin.

Locality.—Claiborne, Ala.

It is impossible for me to decide whether the fragment of a young shell, which H. C. Lea described as *Mitra gracilis*, belongs to this species or not.

CONUS PROTRACTUS, Meyer. Pl. 2, fig. 7. *Am. Jour. Sci.*, XXIX., 1885, p. 466.

Spire nearly a third of the entire length; four smooth, rounded, embryonic whorls are followed by seven carinated volutions; the carina is somewhat crenulated; base with numerous broad, depressed lines.

Localities.—Vicksburg, Miss., Red Bluff, Miss.

Young specimens of *Conus sauridens*, Con., resemble the species, but their spires are concave and covered with revolving lines, which are absent in *C. protractus*.

PLEUROTOMA TEREBRIFORMIS, n. sp. Pl. 2, fig. 8.

Spire high; aperture and canal small, only about a quarter of the entire length; apex blunt, formed by two and a half smooth, embryonic whorls; adult whorls seven or eight—they are flat, have a revolving line above and below and a carina-like, noduliferous double line in the middle; on the older whorls more spirals appear in the interstices; the striæ of growth indicate a flat sinus, situated on the line in the middle; suture distinct.

Localities.—Claiborne, Ala., rare; Newton, Miss., Wheelock, Texas.

The type-specimen is from Claiborne.

PLEUROTOMA JACKSONENSIS, n. sp. Pl. 2, fig. 10.

Spire elevated, rather suddenly decreasing near the blunt apex; aperture and canal about one-third of the entire length; whorls covered with elevated, revolving lines and transverse ribs; a distinctly marked depression on the upper part of each whorl indicates the position of the rather deep sinus; columella thickened; outer lip crenate within.

Locality.—Jackson, Miss., rare.

PLEUROTOMA INFANS, n. sp. Pl. 2, fig. 9.

? *Fusus nanus*, Lea. *Contrib. to Geol.*, p. 150, pl. 5, fig. 155.

? Pleurotoma insignifica, Heilpr. *Proc. Ac. Nat. Sci.*, Philadelphia, 1879, p. 213, pl. 13, fig. 9.

Small; aperture and canal about one-third of the entire length; the pointed apex is formed by two and a half small, smooth, embryonic whorls; three rather large, transversely ribbed, embryonic whorls complete the nucleus; the largest specimen has three adult whorls—they are strongly carinated in the middle; the upper part has only one revolving line near the suture, the lower part three elevated spirals; the upper part indicates the position of the large, regularly rounded sinus; the lines of growth are almost rib-like.

Localities.—Red Bluff, Miss., Newton, Miss., Claiborne? Ala., Vicksburg, Miss (var.).

The type-specimen is from Red Bluff, where the species is not rare; it is much larger than the others of this locality, which have only two adult whorls. The only specimen from Vicksburg, which I have, however, is of the same size, and has also three adult whorls; though otherwise agreeing with the Red Bluff form, it is so decidedly stouter that a varietal name may be properly applied to it—"var. *brevis*." As the same species occurs also in Newton, I have little doubt that it is identical with that Claiborne form which is described by Lea as *Fusus nanus*. If this is so, it is one of the most generally distributed forms of *Pleurotoma* of the Southern Tertiary.

MANGELIA MERIDIONALIS, n. sp. Pl. 1, fig. 14.

Fusiform, somewhat inflated; spire turriculated, acuminated; canal short; nucleus consisting of two rounded and two carinated smooth whorls; adult whorls five, convex, with transverse ribs, about thirteen on each whorl, and with elevated spiral lines; these spirals are small and close together on the uppermost part of each whorl, which is defined by the sinus; sinus broad, near the suture; a curved varix on the outer lip.

Localities.—Claiborne, Ala., Red Bluff, Miss.

From Claiborne I have only the figured type-specimen, without preserved nucleus.

BULLA BITRUNCATA, n. sp. Pl. 2, fig. 24.

Short and stout; spire hidden; columella anteriorly with a strong, oblique fold; surface covered with indistinct, elevated, rounded, revolving lines.

Locality.—Jackson, Miss.

CYLICHNA OVIFORMIS, n. sp. Pl. 2, fig. 32.

Length nearly double the breadth ; umbilicate at the top and with an umbilicate fissure at the columella ; smooth, except some faint impressed, revolving lines.

Localities.—Jackson, Miss., Red Bluff, Miss.

The type-specimen is from Jackson. The form in Red Bluff seems to be a little more flattened.

CYLICHNA JACKSONENSIS, n. sp. Pl. 2, fig. 25.

Length two and a half times the breadth ; umbilicate at the top and with an umbilicate fissure at the columella ; the faint impressed, revolving lines are more distinct near the base and the top.

Locality.—Jackson, Miss.

Considerably more slender than the preceding species. *Cylichna St.- Hilairii,* Lea, sp., from Claiborne, is flat, almost cylindrical, while *Cylichna Jacksonensis* is rounded. The same differences exist in regard to *Cylichna Kellogii,* Gabb, from Texas.

CYLICHNA SUBRADIUS, n sp. Pl. 1, fig. 17.

Cylichna confr. radius, Desh. *Am. Jour. Sci.,* XXIX., 1885, p. 468.

Small, rounded cylindrical ; spire produced, acute ; inferior part, except the base, with impressed, revolving lines ; base with a minute, umbilicate fissure ; aperture but little widening.

Locality.—Claiborne, Ala.

Belongs to the subgenus *Volvula.* In its slender and not inflated form and its acute top, it differs essentially from *Cylichna Dekayi,* Lea, sp., from the same locality. A similar species, however, seems to be *Bulla cylindrus,* H. C. Lea,[*] from the Miocene. It scarcely differs from *Bulla radius,* Desh.,[†] of the Paris basin. A specimen from Wheelock, Texas, determined as *Volvula Conradiana,* Gabb,[‡] is less slender, but else very similar. An examination of a large amount of material may lead to a unison of *Cylichna subradius* with one or all of the last-named species.

ACTÆON ANNECTENS, Meyer. Pl. 2, fig. 30.

Am. Jour. Sci., XXIX., 1885, p. 466.

Conic-ovate ; whorls six ; one and one-half smooth, embryonic whorls

---

[*] *Trans. Am. Philos. Soc.,* 1843, IX., Second Series, p. 250, pl. 35, fig. 43.

[†] Desh., An. s. vertèb., II., p. 226, pl. 39, figs. 22, 23.

[‡] *Jour. Ac. Nat. Sci.,* Philadelphia, IV., Second Series, 1860, p. 386, pl. 67, fig. 51.

form a blunt apex; adult whorls nearly flat, covered with elevated, flat, smooth spiral lines, separated by interstices of about equal breadth. These interstices are set with rather distant, elevated, transverse riblets, which, under the glass, give to the shell an almost cancellated appearance. Inner lip with a very strong and broad fold; in the oldest specimens, with a smaller one above it.

Localities.—Jackson, Miss., Red Bluff, Miss.

The figured specimen from Jackson is much larger than the others; and from Red Bluff I have only the smaller form, represented by more than a dozen specimens. Resembles *Actæon punctatus*, Lea., from Claiborne, but is much smaller, has a stronger fold, besides an additional one. The interstices between the revolving lines are larger, and the small ribs in them more distant and distinct.

ACTÆON INFLATIOR, n. sp.   Pl. 2, fig. 31.

Globose, apex blunt; whorls six, covered with revolving impressed punctate lines; columella with one fold.

Locality.—Claiborne, Ala.

The longitudinal lines are continuous along the volutions; on the body whorl of the type-specimen, however, another system of these lines begins to appear between them. The single fold, together with the globose, regularly rounded form, distinguish this species from the other Actæon of the Southern Tertiary.

PTEROPODA.

STYLIOLA SIMPLEX, n. sp.   Pl. 3, fig. 10.

Shell subulate, nearly straight, smooth; section circular.
Locality.—Jackson, Miss.

The closed end of this species is not inflated.

STYLIOLA HASTATA, n. sp.   Pl. 3, fig. 11.

Shell subulate, nearly straight; section circular; closed end inflated.
Localities.—Vicksburg, Miss. (Higher and Lower Vicksburgian). Red Bluff, Miss.

Seems to be of smaller size than the preceding species. The type-specimen is from the Lower Vicksburgian.

BOVICORNU, n. gen.

Shell minute, subulate pointed, spirally contorted.

BOVICORNU EOCENENSE, n. sp. Pl. 3, fig. 2.

Smooth, somewhat inflated at the closed end ; section circular.
Locality. — Red Bluff, Miss.

LAMELLIBRANCHIATA.

ARCA INORNATA, n. sp. Pl. 1, fig. 24.

Trapezoidal ; anterior side truncated, flat ; beak small ; ligament area very low ; teeth smallest toward the middle ; covered with indistinct concentric lines ; margin entire.
Locality. — Claiborne, Ala.

Resembles *Arca lævigata*, Caillat,* from the Paris basin, but is less oblong.

TRIGONOCŒLIA LEDOIDES, n. sp. Pl. 1, fig. 20.

Convex, ovate ; posterior side carinated ; hinge narrow, divided by a minute pit ; about eight teeth on each side ; more vertical near the pit, more horizontal near the end ; surface with concentric lines of growth crossed by indistinct radiating lines, which are not perceptible on the umbo ; beak turned toward the carinated side ; the muscular impression of this side is oval and less distinct ; an elevated radiating line passes along its side ; margin entire.
Locality. — Claiborne, Ala.

I found only the figured specimen, and this seems to be somewhat worn.

LEDA MATER, Meyer. Pl. 3, fig. 20. *Am. Jour. Sci.*, XXIX., 1885, p. 460.

Elliptically transverse, convex ; produced and truncate behind ; strongly inequilateral ; covered with concentric ribs and posteriorly with three radiating ribs ; teeth diminishing in size toward the fosset ; channel interrupted by a callus ; margin entire.
Locality. — Jackson, Miss. Not rare.

The concentric ribs are quite regular on the umbonial part of the shell :

---

* Desh. An. s. vertéb. I., p. 905, pl. 68, figs. 23–26.

on the ventral part, however, about half of them vanish rather suddenly posteriorly, and the rest of them increase in size, so that the posterior part is covered by fewer but larger ribs. Many of the specimens have one or two indistinct radiating furrows on the anterior part. In some, the concentric ribs become erased like on the anterior part (see *loc. cit*).

LEDA TRIANGULATA, n. sp.　Pl. 3, fig. 14.

Triangular; ventral margin rounded; near equilateral; ventricose; hinge-plate long and broad; surface nearly smooth, with indistinct concentric lines; margin entire.

Locality.—Red Bluff, Miss.

I found only the figured specimen.

LIMOPSIS RADIATUS, Meyer.　Pl. 3, figs. 17, 17a.　*Am. Jour. Sci.*, XXIX., 1885, p. 459.

Rounded, quadrangular; solid; hinge teeth, diminishing in size near the pit; surface covered by alternating radiating ribs, crossed by equal closely set, elevated concentric lines; the points of crossing are thickened by nodules; margin crenulate within.

Locality.—Jackson, Miss.　Common.

The similarity of this species with *Lymopsis obliquus*, Lea, sp., is pointed out in the above cited place.

ASTARTE PROTRACTA, n. sp.　Pl. 3, figs. 18, 18a.

Elongated, subquadrangular; umbonial part strongly flattened; adductors prominent; surface covered with distant concentric ribs, becoming obsolete on the umbonial part; margin entire.

Locality.—Enterprise, Miss.

The single specimen was collected by me in the stratum, in Enterprise, mentioned *Am. Jour. Sci.*, XXX., 1885, p. 70, as upper bed.

ASTARTE TRIANGULATA, n. sp.　Pl. 3, figs. 21, 21a.

Trigonal, solid; pedal scar of anterior adductor distinct; lunule long and flat; surface closely covered with concentric ribs; margin crenulate.

Locality.—Red Bluff, Miss.　Common.

The concentric ribs vary in size, in different specimens, and in some become obsolete toward the ventral margin

MICROMERIS SENEX, n. sp. Pl. 3, fig. 22.

Small, solid; anterior margin straight, posterior margin curved; surface, except the umbo, covered with coarse radiating ribs.

Locality.—Claiborne, Ala. Stratum "g".*

LUCINA (CYCLAS) SUBRIGAULTIANA, n. sp. Pl. 3, fig. 13, 13a.

Orbicular, very regularly rounded; hinge teeth, two; laterals small; margin entire; surface covered with regular concentric waving lines, crossed occasionally by orbicular lines of growth.

Locality.—Vicksburg, Miss. Lower Vicksburgian.

Resembles very much *Lucina rigaultiana*, Desh.† A deep but extremely small lunule is in front of the beak. The lines of ornamentation are less distinct at their convex summit, thus creating a somewhat erased zone, which radiates from the umbo toward the posterior side of the ventral margin. The ornamentation looks as if produced by "*petits plans glissant les uns au dessous des autres.*"

LUCINA CHOCTAVENSIS, n. sp. Pl. 1, fig. 28.

Small, suborbicular; convex; lunule semilunar, well defined; cardinal and lateral teeth; anterior muscular impression relatively small; covered with concentric lines of growth.

Locality.—Vicksburg, Miss.

Lucina papyracea, Lea, from Claiborne, is similar, but the lateral teeth are obsolete.

LUCINA SMITHII, n. sp. Pl. 1, fig, 23.

Solid; irregularly elliptical; subequilateral; with cardinal and lateral teeth; a small but deep lunule inside of a larger indistinct one; surface with irregular lines of growth, becoming more regular at the extremities; they are crossed by very indistinct radiating lines; margin crenulate.

Locality.—Claiborne, Ala.

Named in honor of Dr. E. A. Smith. Tuscaloosa, Ala.

LUCINA BISCULPTA, n. sp. Pl. 1, figs. 30, 30a.

Thin; convex; oval; anterior margin truncated; beak turned anteriorly; lunule rather small; impressed; cordate; hinge of the right valve

---

* Am. Jour. Sci., XXX., 1885., p. 69.
† Desh. An. s. vertèb. I., p. 631, pl. 47, figs. 28–30.

with one cardinal and two obsolete distant lateral teeth ; inner surface with radiating impressed lines, which crenulate the margin ; outer surface with concentric elevated distinct lines ; except on the umbo they are separated by rather large and regular distances.

Locality.—Claiborne, Ala.

The radiating lines of the inside correspond, at some places near the margin, with indistinct rib-like elevations of the surface.   By a mistake, the figure of this species, on plate 1, shows two teeth below the beak instead of one.

MACTRA INÆQUILATERALIS, n. sp.   Pl. 1, fig. 18.

Substance of the shell thick ; inæquilateral ; triangular ; anteriorly rounded ; posterior slope carinated ; covered with indistinct concentric lines, which are distinct on the extremities.

Locality.—Vicksburg, Miss.

As Conrad describes and figures *Mactra funerata** as equilateral, I am unable to refer the figured specimen to his species.

HINDSIELLA FABA, n. sp.   Pl. 1, fig. 25.

Small, convex ; sinus of the ventral margin gently rounded ; covered with irregular lines of growth ; muscular impressions ovately elongated ; an indistinct oblique tooth below the beak (right valve).

Locality.—Claiborne, Ala.

Allied to *Hindsiella arcuata*, Lam., sp.† of the Paris basin.

ERYCINA WHITFIELDI, n. sp.   Pl. 1, fig. 29.

Small ; substance of the shell thin, porcellaneously shining ; oval ; convex ; inæquilateral, the anterior side the largest ; left valve with two obsolete pyramidal teeth on the anterior side, and on the posterior side a compressed one, which is very small and distant from the beak ; surface shining, smooth, except very fine concentric lines of growth ; margin entire.

Locality.—Claiborne, Ala.

The nearest species in the Paris basin seems to be *Erycina obsoleta*, Desh.‡   Named after Prof. R. P. Whitfield, whose descriptions and figures of American Eocene shells belong to the most careful ones.

---

* *Jour. Ac. Nat. Sci.*, I., 2d Series, 1848, p. 121, pl. 12, fig. 13.

† Desh., An. s. vertèb., I., p. 695, pl. 53, figs. 32-35.

‡ Desh., An. s. vertèb., I., p. 720, pl. 53, figs. 16-19.

KELLIELLA? BŒTTGERI, n. sp.   Pl. 3, figs. 15, 15a.

Very small, orbicular, tumid, inæquilateral, umbo turned anteriorly ; a cordate lunule is defined by an impressed line ; surface closely and regularly covered with concentric ribs ; hinge of the right valve with two diverging cardinal teeth below the umbo, and a horizontal one which is lamelliform before and beneath them ; left valve with a short, oblique tooth below the umbo, and anteriorly with a horizontal S-shaped one ; anterior adductor long ; margin entire.

Locality.—Jackson, Miss.; common.

The pallial line is apparently simple. The shell in its dentition resembles somewhat the genus *Lutetia,* Desh., of the Paris basin, but on comparison with two species of this genus which I have, proves to be different. The genus *Kelliella,* Sars., as far as I am aware, is not a known fossil. I have no specimens of *Kelliella,* but according to its description and figure, I am inclined to put this small, remarkable Jackson shell into this genus.

MODIOLARIA ALABAMENSIS, n. sp.   Pl. 3, fig. 19.

Rhomboidal, thin ; the small anterior and the large posterior part with radiating ribs, leaving the middle of the shell and the umbo smooth ; hinge edentulous, anterior hinge-line notched.

Locality.—Claiborne, Ala.; "Lowest Claibornian."

The figured type-specimen is a young shell.

CORBULA ALDRICHI, Meyer.   Pl. 1, fig. 21. *Am. Jour. Sci.*, XXX., 1885, p. 67.

Rounded trigonal ; ventricose ; posterior side carinated ; beak small, curved anteriorly, in the left valve nearly in the middle ; right valve briefly rostrated ; in both valves the umbonial part is without concentric ribs, but with impressed, radiating lines—the ventral part with concentric ribs.

Locality.—Wood's Bluff, Ala.

The radiating lines cut only the first ribs and disappear completely at the ventral part. The species is similar to *Corbula gibbosa,* Lea., but distinguished mainly by the smooth umbonial part and the radiating lines.

CORBULA PEARLENSIS, n. sp.   Pl. 3, figs. 16, 16a.

Rather small, rounded, inflated ; margin rounded anteriorly, truncated posteriorly ; beaks very small, turned anteriorly ; surface of both valves smooth on the umbonial part, covered with rounded concentric ribs on the ventral part.

Locality.—Jackson, Miss., rare.

VENERICARDIA INFLATIOR, Meyer.   Pl. 1, fig. 26.   *Am. Jour. Sci.*, XXIX., 1885, p. 460.

Small ; cordate ; ventricose ; subæquilateral ; margin subcircular ; beak elevated and large, turned anteriorly ; surface covered with about twenty ribs ; they are smooth at the umbo, slightly crenulated toward the ventral margin ; the interstices are about of the same size as the ribs ; margin crenulated.

Locality.—Claiborne, Ala.

Differs from *Venericardia parva, Lea.*, from the same locality, principally in being smaller, more ventricose and rounded, and having a much larger beak.

VENUS RETISCULPTA, n. sp.   Pl. 1, figs. 27, 27a.

Rhombically circular ; convex ; covered with broad, flat, and somewhat irregular concentric lines, crossed by similar radiating ones ; the sculpture indistinct near the umbo ; an impressed line separates a long lunule ; pallial sinus large ; margin entire.

Locality.—Claiborne, Ala.

The sculpture makes the surface appear as if reticulated with small rounded holes, especially in worn shells ; the figured specimen shows the complete hinge, otherwise it is not the largest or best sculptured one.

ALVEINUS MINUTUS, Conrad.   Pl. 1, fig. 19.

Alveinus parva, Con., *Am. Jour. Conch.*, 1865, p. 10.—List name.
Alveinus minuta, Con., *Am. Jour. Conch.*, 1865, p. 138.   Pl. 10, fig. 2.
Alveinus parvus, Con., *Check List of Tertiary invertebr.*, 1866, p. 24.
Alveinus minutus, Con., *Proc. Acad. Nat. Sci.*, Phila., 1872, p. 53. Pl. 1, fig. 6.
Alveinus minutus, Con., *Am. Jour. Sci.*, XXIX., 1885, p. 467.

Small, nearly circular ; inæquilateral ; surface with concentric lines of growth, otherwise smooth and somewhat shining ; beaks small, turned anteriorly ; hinge divided by a minute trigonal pit ; right valve with a pyramidal tooth on the anterior side ; left valve with two compressed ones ; muscular impressions subquadrangular, of about equal size ; margin microscopically channeled within ; pallial line simple.

Conrad described this species with the name of locality, "Enterprise, Miss." The specimen figured on plate 1 is from Claiborne, Ala., where the species is not rare. It is common in Jackson, Miss., and altogether occurs in so many other localities, that it may be considered as characteristic for the Southwestern Tertiary.

PERIPLOMA COMPLICATA, n. sp.   Pl. 1, fig. 22.

The only specimen found is of the same fragmentary shape as that in which *Periploma Claibornensis*, Lea., sp., a species not rare, constantly occurs. The fissure of the beak and the pearly nacre of the inside are well seen ; the hinge with the spoon-shaped process is much more complicated than in *P. Claibornensis*, as the figure shows, the process consisting mainly of two concentric spoons.

Locality.—Claiborne, Ala.

ECHINODERMATA.

ECHINOCYAMUS HUXLEYANUS, n. sp.   Pl. 3, fig. 23.

The only specimen found is fragile and damaged ; it is egg-shaped, depressed, with rather coarse tubercles, large mouth and vent, the latter near the margin : a similar species is *Echinocyamus oviformis*, Forbes.[*]

Locality.—Claiborne, Ala.

FORAMINIFERA.

NODOSARIA OBLIQUA, L. sp.   Pl. 1, fig. 31.

Of this widespread species I give a figure, as it is the only specimen of Foraminifera found as yet in the Claiborne Sand. The species occurs in most of the many localities of the Southwestern Tertiary

---

[*] *Pub. Pal. Soc.*, London.

Explanation of Plate I.

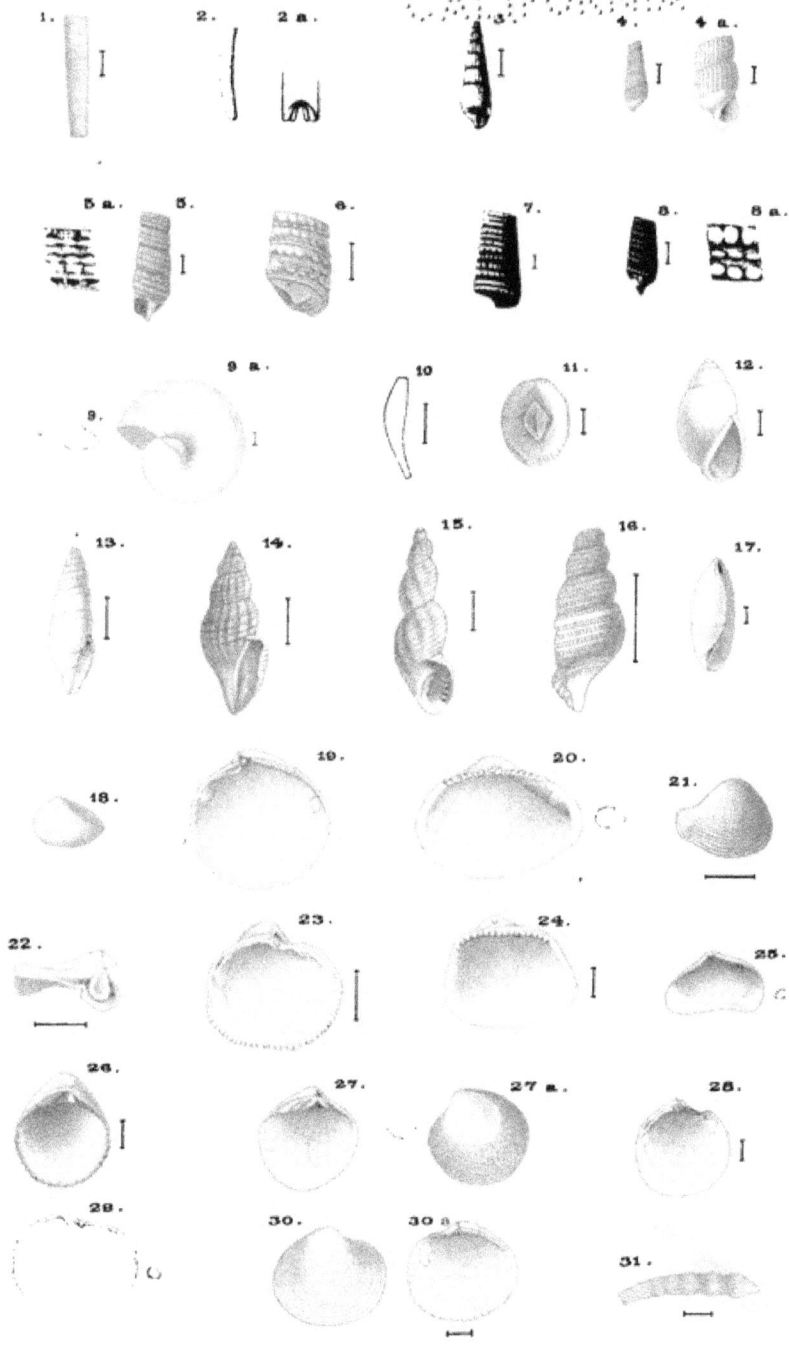

CALIFORNIA

Explanation of Plate II.

CALIFORNIA

2   3   4   5   6

8   9   10   11

   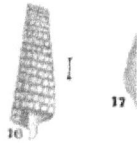

13   14   15   16   17

19   20a   20   21

 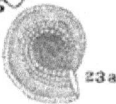

23b

23   23a   24   2

27

27a

29   29a   30   31   32

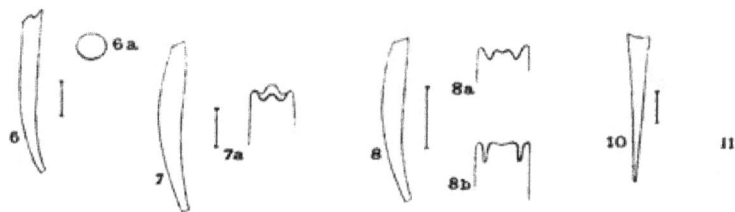

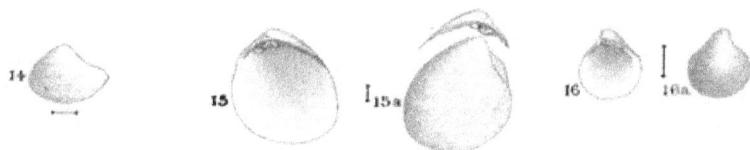